P9-EDK-058

Functional Diversity of Plants in the Sea and on Land

A. R. O. Chapman
Dalhousie University
Halifax, Nova Scotia, Canada

Illustrations by Pat Evans-Lindley

Jones and Bartlett Publishers, Inc.
Boston/Portola Valley

Cover Art: From interior. Figure 6: Common microscopic genera of marine phytoplankton.

Modified after Round, F. E., 1981. *The Ecology of Algae.* Cambridge University Press, Cambridge.

Editorial offices: 30 Granada Court, Portola Valley, CA 94025

Sales and customer service offices: 20 Park Plaza, Boston, MA 02116

Library of Congress Cataloging-in-Publication Data

Chapman, A. R. O.
Functional diversity of plants in the sea and on land.

Includes index.
1. Botany—Variation. 2. Botany—Ecology. 3. Botany—Classification. 4. Plants—Evolution. I. Title.
QK983.C48 1987 581.5 85-24129
ISBN 0-86720-064-2

Production: Unicorn Production Services, Inc.
Design: Erika Petersson
Compositor: WordTech Corporation
Printer and Binder: Halliday Lithograph

Printed in the United States of America

Printing number (last digit) 10 9 8 7 6 5 4 3 2 1

To Jan, Paul, Lin and Hanah

Preface

This book is intended to introduce first and second year university students to major groups of plants within the context of the environments that they live in. Most of the plants that live in the sea are algae, and most of the plants on land are vascular plants. It is, therefore, possible to present systematic diversity of algae with reference to the aquatic mode of life. Similarly, vascular plant diversity can be presented in terms of the problems of life on land. My experience over more than ten years of teaching a biological diversity course is that students find this approach inherently appealing. The students learn about plant diversity, but they also learn about how plants function in diverse habitats.

I am especially grateful to Pat Evans-Lindley who prepared all of the illustrations. I would also like to thank all of the authors and publishers who have allowed me to use their illustrations. Pat Harding provided an in-depth critique of the manuscript which made me explain some of my ideas much more clearly. I am very grateful to her. The major part of the manuscript was prepared during a sabbatical year that I spent at the University of Bristol. I would like to thank Professor Frank Round for making arrangements for my stay and also for his permission to use the illustration shown in Figure 6 and on the cover of this book.

A.R.O. Chapman
Dalhousie University, Canada
September 1985

Contents

Figure Sources

Most of the text figures were modified by Pat Evans-Lindley from the sources shown below.

Figure 2. Tappan, H. N. 1980. *The Palaeobiology of Plant Protists*. W. H. Freeman, San Francisco.

Figures 3, 4. Mann, K. H. 1982. *Ecology of Coastal Waters. A Systems Approach*. Blackwell, Oxford.

Figure 5. Neushul, M. 1972. Functional interpretation of benthic algal morphology. In *Contributions to the Systematics of Benthic Marine Algae of the North Pacific*, eds. I. A. Abbott and M. Kurogi, 47–73. Japanese Society of Phycology, Kobe.

Figure 6. Round, F. E. 1981. *The Ecology of Algae*. Cambridge University Press, Cambridge.

Figure 7. Margalef, R. 1978. *Oceanologia Acta*. 1: 493–509.

Figure 8. Reynolds, C. S. 1984. *The Ecology of Freshwater Phytoplankton*. Cambridge University Press, Cambridge.

Figure 9. Margulis, L. 1982. *Early Life*. Jones and Bartlett Publishers, Boston.

Figure 11. Syrrett, P. J. 1981. Nitrogen metabolism of microalgae. *Can. Bull. Fish. Aquat. Sci.* 210: 182–210.

Figure 12. Haselkorn, R. 1978. *Ann. Rev. Plant Physiol.* 29, 319–344.

Figures 13, 16, 19, 21, 22, 25, 27, 35, 42. Lee, R. E. 1980. *Phycology*. Cambridge University Press, Cambridge.

Figures 14, 17, 23, 24, 28, 31, 37, 38, 39, 40. Scagel, R. F., R. J. Bandoni, G. E. Rouse, W. B. Schofield, J. R. Stein and T. M. C. Taylor. 1965. *Plant Diversity: An Evolutionary Survey*. Wadsworth, Belmont.

Figure 18. Scagel, R. F., R. J. Bandoni, G. E. Rouse, W. B. Schofield, J. R. Stein and T. M. C. Taylor. 1965. *An Evolutionary Survey of the Plant Kingdom*. Wadsworth, Belmont.

Figure 20. Esser, K. 1982. *Cryptogams*. Cambridge University Press, Cambridge.

Figure 26. Rabenhorst, L. 1933. *Kryptogamen-Flora*. Vol. X, Section

III, Part I. *Dinoflagellatae.* Akademische Verlagsgesellschaft M. B. H., Leipzig. Fott, B. 1959. *Algenkunde.* VEB Gustav Fischer, Jena.

Figures 28, 29. Littler, M. M., D. S. Littler and P. R. Taylor. 1983. *J. Phycol.* 19: 229–237.

Figure 30. Koehl, M. A. R. 1982. *Sci. Amer.* 247: 124–134.

Figures 31, 32, 34. Chapman, A. R. O. 1979. *Biology of Seaweeds: Levels of Organization.* University Park Press, Baltimore.

Figure 33. Nicholson, N. 1976. *Bot. Mar.* 19: 23–31.

Figure 36. Bouck, G.B. 1965. *J. Cell. Biol.:* 523-537.

Figure 40, 43. Hoek, C. van den. 1981. Chlorophyta: Morphology and classification. In *The Biology of Seaweeds,* eds. Lobban, C. S. and M. Wynne, 86–132. Blackwell, Oxford.

Figure 44. Womersley, H. B. S. 1970. *Phycologia* 10: 229–233.

Figure 45. McNeill Alexander, R. 1971. *Size and Shape.* Edward Arnold, London.

Figures 46, 48, 54. Wainwright, S. A., W. D. Biggs, J. D. Currey and J. M. Gosline. 1976. *Mechanical Design of Organisms.* Princeton University Press, Princeton.

Figures 49, 58, 59, 72. Haberlandt, G. 1914. *Physiological Plant Anatomy.* MacMillan, London.

Figure 52. Mark, R. E. 1967. *Cell Wall Mechanics of Tracheids.* Yale University Press, New Haven.

Figure 53. Esau, K. 1976. *Anatomy of Seed Plants.* Wiley, New York.

Figures 55, 68. Skene, M. 1924. *The Biology of Flowering Plants.* Sidgwick and Jackson, London.

Figures 60, 65. Larcher, W. 1975. *Physiological Plant Ecology.* Springer-Verlag, Berlin.

Figures 61, 62, 67. Leopold, A. C. and P. E. Kriedmann. 1975. *Plant Growth and Development.* McGraw-Hill, London.

Figure 63. Coult, D. A. 1973. *The Working Plant.* Longman, London.

Figure 64. Barbour, M. G., J. H. Burke and W. D. Pitts. 1980. *Terrestrial Plant Ecology.* Benjamin Cummings, Menlo Park.

Figure 69. Parker, J. 1968. Drought resistance mechanisms. In *Water Deficits and Plant Growth. Vol. 1. Development, Control and Measurement,* ed. T. T. Kozlowski, 195–234. Academic Press, New York.

Figures 70, 71. Goebel, K. 1889. *Pflanzenbiologische Schilderungen.* N. G. Elwert'sche, Marburg.

Figure 73. Gessner, F. 1956. Wasserspeicherung und Wasserverscheibung. In *Encyclopaedia of Plant Physiology. Vol. III. Water Relations of Plants*, ed. W. Ruhland, 247–256. Springer-Verlag, Berlin.

Figures 75, 79, 92, 93, 98, 103. Foster, A. S. and E. M. Gifford. 1974. *Comparative Morphology of Vascular Plants.* W. H. Freeman, San Francisco.

Figure 76. Meyer, B. S. and D. B. Anderson. 1939. *Plant Physiology.* Litton Educational Publishing Co., New York.

Figure 77. Brown, W. H. 1935. *The Plant Kingdom.* Ginn and Co., Lexington.

Figure 81. Delevoryas, T. 1966. *Plant Diversification.* Holt, Rinehart and Winston, New York.

Figures 82, 85. Savidge, J. P. 1976. The angiosperm flower and related structures. In *Plant Structure, Function and Adaptation*, ed. Hall, M. A., 326–378. MacMillan, London.

Figure 83. Canright, J. E. 1952. *Am. J. Bot.* 39: 484–497.

Figure 84. Sinnott, E. W. and K. S. Wilson. 1963. *Botany, Principles and Problems.* McGraw-Hill, New York.

Figure 86. Robbins, W. W., T. E. Weier and C. R. Stocking. 1964. *Botany.* Wiley, New York.

Figures 87, 88. Grant, V. 1963. *The Origin of Adaptations.* Columbia University Press, New York.

Figure 89. Leppik, E. E. 1957. *Evolution* 11: 466–481.

Figure 90. Kidston, R. and W. H. Lang. 1921. *Trans. Roy. Soc. Edin.* 52, Part IV.

Figure 91. Mägdefrau, K. 1953. *Paläobiologie der Pflanzen.* G. F. Fischer, Jena.

Figures 92, 96, 98, 101, 102. Stewart, W. N. 1983. *Paleobotany.* Cambridge University Press, Cambridge.

Figure 93. Andrews, H. N., A. Kaspar and E. Mencher. 1968. *Bull. Torrey Bot. Club.* 95:1–11.

Figure 94. Jeffrey, E. C. 1917. *The Anatomy of Woody Plants.* University of Chicago Press, Chicago.

Figure 96. Bruchmann, H. 1912. *Flora.* 104: 180–224.

Figure 99. Emberger, L. 1968. *Les Plantes Fossiles.* Masson et Cie, Paris.

Figure 104. Maheshwari, K. 1935. *Proc. Indian Acad. Sci.* 1: 586–606.

Figures 106, 108, 110, 111, 112, 113. Hébant, C. 1977. *The Conducting Tissues of Bryophytes.* J. Cramer, Vaduz.

Figure 107. Mägdefrau, K. 1982. Life forms of bryophytes. In *Bryophyte Ecology*, ed. Smith, A. J. E., 45–58. Chapman and Hall, London.

Figures 109, 114, 115, 117, 118, 119, 120, 122. Parihar, N. S. 1961. *An Introduction to the Embryophyta. I. Bryophyta.* Cental Book Depot, Allahabad.

Figure 116. Smith, G. M. 1955. *Cryptogamic Botany. Vol. II.* McGraw-Hill, New York.

Figure 121. Ingold, C. T. 1965. *Spore Liberation.* Clarendon Press, Oxford.

Figures 123, 135. Burnett, J. H. 1976. *Fundamentals of Mycology.* Edward Arnold, London.

Figure 124. Cole, G. T. 1981. Architecture and chemistry of the cell walls of higher fungi. In *Microbiology*, ed. D. Schlessinger. American Society for Microbiology, Washington, D. C., 227–231.

Figures 136, 144, 149, 152. Pritchard, H. N. and P. T. Bradt. 1984. *Biology of Nonvascular Plants.* Times Mirror/Mosby, St. Louis.

Figure 125. Neushul, M. 1974. *Botany.* Hamilton, Santa Barbara.

Figures 126, 129, 131, 138. Ingold, C. T. 1961. *The Biology of Fungi.* Huchinson, London.

Figures 128, 140, 141. Alexopoulos, C. J. 1962. *Introductory Mycology.* Wiley, New York.

Figure 130. Hale, M. E. 1967. *The Biology of Lichens.* Edward Arnold, London.

Figures 132, 134, 139. Ingold, C. T. 1971. *Fungal Spores. Their Liberation and Dispersal.* Clarendon Press, Oxford.

Figures 146, 148, 150. Scagel, R. F., R. J. Bandoni, J. R. Maze, G. E. Rouse, W. B. Schofield and J. R. Stein. 1982. *Nonvascular Plants. An Evolutionary Survey.* Wadsworth, Belmont.

Figure 151. Garrett, S. D. 1963. *Soil Fungi and Soil Fertility.* Pergamon Press, Oxford.

1

Introduction

Botanical diversity encompasses all photosynthetic plants and all of the fungi. The diversity within and among these groups is overwhelming and difficult to systematize taxonomically or phylogenetically. However, it is necessary to review the taxonomic and historical relationships of plant groups before embarking on a consideration of the functional relationships between plants and their habitats (which is what this book is about). In this introduction a short and elementary treatment of plant systematics will be presented from a phylogenetic perspective.

Plant Taxa

The diversity of living organisms is classified in a taxonomic hierarchy. The base point of this hierarchy is of the myriad **species** in the world. Species mean different things to different people, but the **morphological species** concept seems most appropriate in the present context. Members of a morphological species are similar to one another in most respects and have correlated morphological characteristics which are not shared with members of other species. Groups of species are agglomerated in the next highest category known as the **genus** (Fig. 1). Genera are grouped into **families**, families into **orders**,

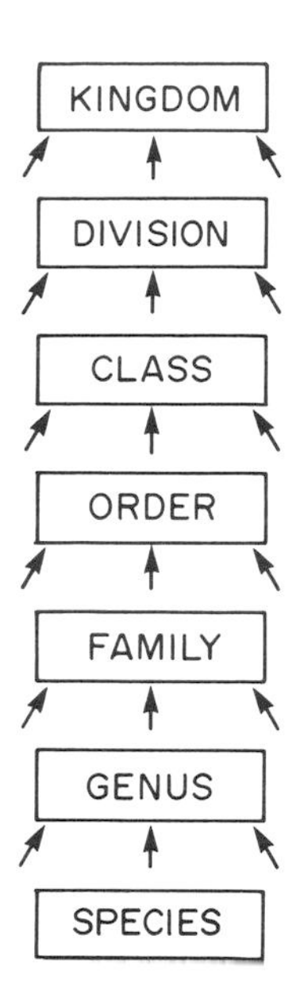

Figure 1. Hierarchy of taxonomic categories. Taxa at lower levels are grouped into higher level taxa.

orders into **divisions**, and divisions into **kingdoms**.

There is considerable controversy over the way in which kingdoms of living organisms are to be separated from one another. Plants and Animals are often regarded as separate kingdoms, but although antelopes and daffodils are clearly easy to differentiate, the separation of single celled plants from single celled animals is well nigh impossible. In fact, diversity in the biological world falls clearly into two kingdoms known as the **Prokaryota** and the **Eukaryota**. The differences between these two groups are tabulated in Table 1. In Table 2 the divisions of living organisms considered in this book are partitioned between the Prokaryota and the Eukaryota. There is no confusion. Each division is either prokaryotic or eukaryotic and thus we recognize two kingoms of living organisms.

Where do we find botanical diversity among the two kingdoms? This is not an easy question to answer. Botany has traditionally comprised a study of the groups listed in Table 2 (bacteria other than Cyanobacteria are not included here, but are considered in some botanical texts). In colloquial terminology these groups are known as **blue-green algae**, **algae**, **bryophytes**, **vascular plants** and **fungi**. The blue-green algae and the algae are mostly aquatic; bryophytes, vascular plants and fungi are mostly terrestrial. The aquatic and terrestrial modes of life comprise the theme of this treatment of botanical diversity. However, the historical development of the modes of life must first be considered.

Table 1. Some important differences between cell types in the Kingdoms Prokaryota and Eukaryota.

Characteristic	*Prokaryota*	*Eukaryota*
Nuclear membrane	Absent	Present
Chromosomes	Composed of nucleic acid only	Composed of nucleic acid and protein
Cytoplasmic organelles	Absent	Present
Flagella	Lack 9+2 fibril organization	Have 9+2 fibril organization
Cell wall	Contains peptidoglycans as supporting polymers*	Does not contain peptidoglycans

* Peptidoglycans are built up of N-substituted glucosamines and muramic acid (3-0 lactylglucosamine). The archaebacteria have a pseudo-peptidoglycan constructed on the same general plan, but with different chemical components.

Table 2. Distribution of divisions (dealt with in text) among two kingdoms.

Kingdom	*Division*	*Common Name*
Prokaryota	Cyanobacteria	Blue-green algae
Eukaryota	Rhodophyta	Red algae
	Chlorophyta	Green algae
	Euglenophyta	Euglenids
	Chrysophyta	Golden algae
	Pyrrophyta	Dinoflagellates
	Cryptophyta	Cryptomonads
	Phaeophyta	Brown algae
	Tracheophyta	Vascular plants
	Bryophyta	Mosses, liverworts, hornworts
	Chytridiomycota	Water molds, chytrids
	Oomycota	Water molds, downy mildews
	Zygomycota	Pin molds
	Ascomycota	Sac fungi
	Basidiomycota	Club fungi
	Deuteromycota	Imperfect fungi

History of Plant Life

The earth is about 4.5 billion years old. When it first condensed there was probably no atmosphere, and no life. The first atmospheric gases came from volcanic activity and probably consisted of CH_4, NH_3, H_2 and H_2O vapor. Free oxygen was almost certainly absent from this primitive atmosphere.

The gases in the atmosphere of the primeval world would be noxious to most modern life forms. Furthermore, for most of its history the earth was bombarded with lethal ultra-violet radiation (UV) from the sun. Life on earth is now protected from this radiation by the layer of ozone gas in the high atmosphere. The ozone is derived from the oxygen produced in the lower atmosphere by plant photosynthesis. The development of an oxygen rich atmosphere has been a very slow process. Because of UV radiation early life was confined to aquatic habitats where water screens out the harmful rays. Initially the radiation was harmful to perhaps 10 meters depth. Presumably life was then restricted to deeper water. Since water rapidly attenuates light in the visible range (as well as UV), photosynthetic production must also have been restricted by inadequate illumination. Because the rate of photosynthesis determines the release of oxygen into the atmosphere, the development of an oxidizing atmosphere must therefore have been very slow. Oxygen in the atmosphere reached 1% of present levels about 600-650 million years ago (MYA). It took perhaps a billion years of photosynthesis for this level to be achieved.

Who were the first photosynthesizers and who were their ancestors? There is some consensus in the view that the first living organisms on earth were not photosynthetic or autotrophic in any respect. Most probably they were fermenting heterotrophs. All of these terms need explanation, and it will be necessary now to embark on a discussion of the ways in which organisms obtain energy .

Organisms obtain some or all of their energy by the oxidation of an organic compound:

$$AH_2 + B \rightarrow A + BH_2 + \text{Energy}$$

Organic hydrogen donor — Hydrogen acceptor

If the molecule B (hydrogen acceptor) is oxygen, then the process is aerobic respiration, and BH_2 is water. If molecule B is organic, then the process is fermentation and BH_2 is organic:

$$AH_2 \cdot B \rightarrow A + BH_2 + \text{Energy}$$

Here B is part of the substrate.

In fermentation there is no requirement for free oxygen in order to obtain energy from organic substrates. Fermentation is thought to have been the nutritional mode of the first living organisms on earth.

Fermenting anaerobic bacteria are common today. Two common groups are called **lactic acid bacteria** and **clostridia** and these are thought to be very primitive in their life styles. Lactic acid bacteria sour milk and ripen cheeses. Clostridia belong to the genus *Clostridium* and are found in soil, dust, water and animal guts wherever oxygen is absent and an organic substrate is present.

Apart from fermentation, anaerobic respiration can occur when the hydrogen acceptor is an oxidized inorganic substance like sulfate:

$$4AH_2 + H_2SO_4 \rightarrow 4A + H_2S + 4H_2O + \text{Energy}$$

This is called sulfate respiration. *Desulfovibrio* is a modern sulfate respirer (Fig. 2) that lives in oxygen free sediments containing sulfate and organic matter. The organisms ferment organic compounds to acetic acid and they convert sulfate (SO_4^{2-}) into sulfide (S^{2-}) which may be released as hydrogen sulfide (H_2S) or dimethyl sulfide (H_3CSCH_3).

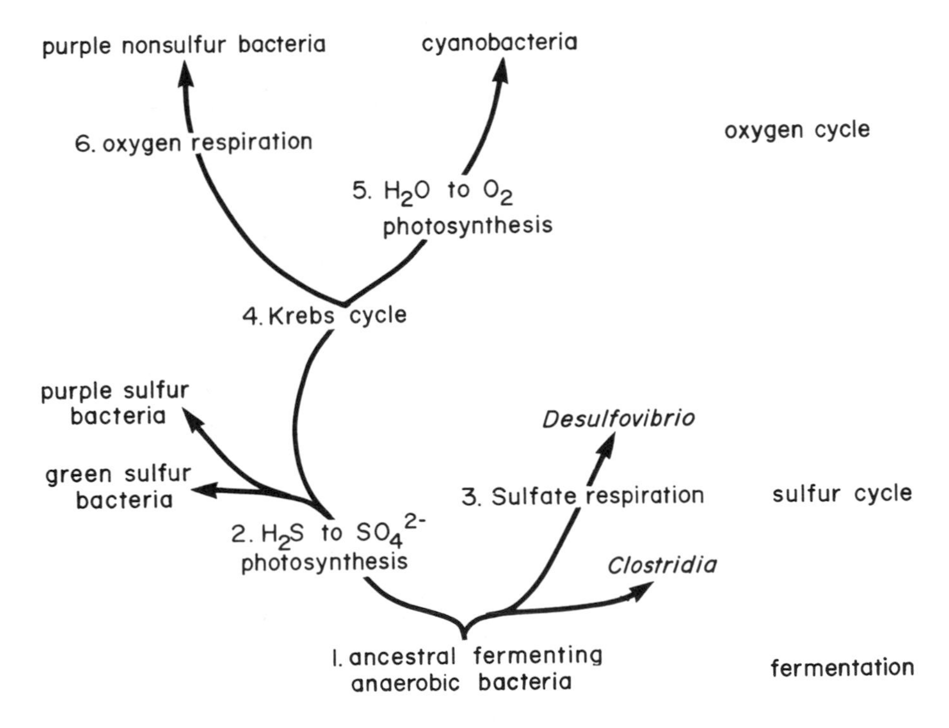

Figure 2. A hypothetical evolutionary diversification of prokaryotes leading to Cyanobacteria and the development of an oxygen-rich atmosphere (modified after Tappan, 1980).

When the organic compound is oxidized in sulfate respiration, hydrogen atoms (or electrons) are transferred to sulfate via a chain of chemicals called the **electron transport chain**. The **cytochromes** (proteins) are major components of the electron transport chain. The electron carrying capacity is conferred by a ring shaped compound called a porphyrin. At the center of a cytochrome porphyrin ring is an iron atom and the porphyrin-metal complex is called a **heme** group. The structure is as follows:

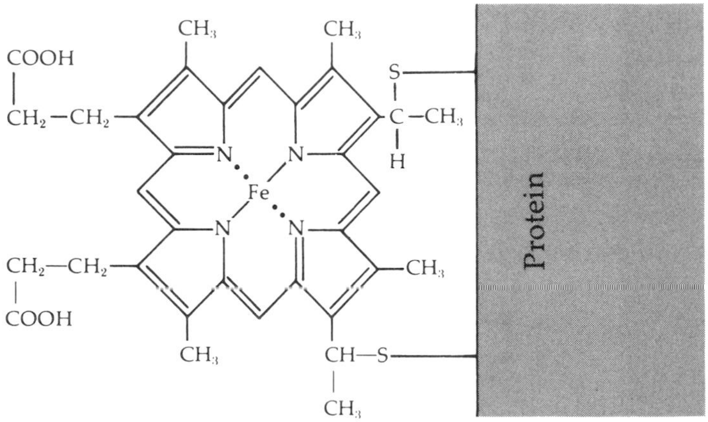

Sulfate respirers may have arisen from fermenting anaerobes, and more importantly, organisms using sulfate and nitrate as hydrogen acceptors may have become photosynthetic by trapping photons when a heme protein mutated to form chlorophyll. Chlorophyll is a porphyrin – magnesium complex. The structure is as follows:

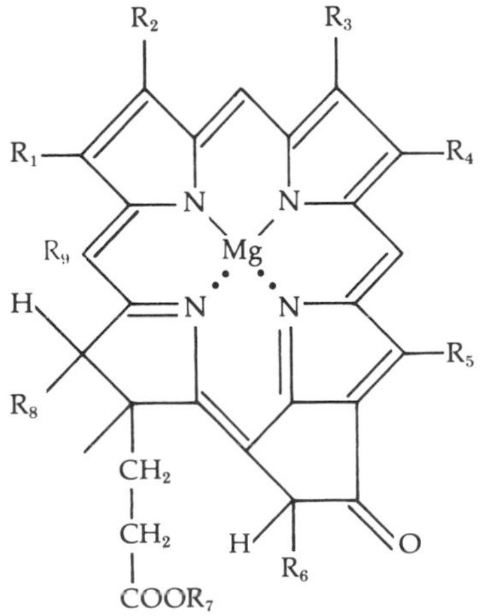

The phylogenetic development of these systems leading to the accumulation of oxygen in the atmosphere are shown in Fig. 2. The important component of this system is the development of photosynthesis in which water is the hydrogen donor. What does this mean? The basic equation for photosynthesis is:

$$CO_2 + H_2A + \text{light} \rightarrow (CH_2O) + H_2O + 2A$$

hydrogen acceptor   hydrogen donor

It is believed that the first photosynthesizers used H_2S as the hydrogen donor. Indeed, photosynthetic bacteria still do this, and so do a number of blue-green algae:

$$2H_2S + CO_2 + \text{light} \rightarrow 2S + H_2O + (CH_2O)$$

In this system sulfur deposits are formed. The green and purple sulfur bacteria (Fig. 2) both take their hydrogen atoms from volcanically or biologically produced hydrogen sulfide and generate sulfur. On the other hand, the purple nonsulfur bacteria (Fig. 2) remove hydrogen atoms from organic compounds such as ethanol or lactic acid, or from hydrogen gas itself. Both types of purple bacteria harvest light energy with bacteriochlorophyll. The green sulfur bacteria have a chlorophyll called **chlorobium**.

When water became the major hydrogen donor the result was a release of free oxygen (ultimately into the reducing atmosphere):

$$2H_2O + CO_2 + \text{light} \rightarrow O_2 + H_2O + (CH_2O)$$

Most blue-green algae use this latter reaction, and the development of these organisms in the aquatic systems of the world almost two billion years ago was a necessary prerequisite for the colonization of land and of shallow water by the eukaryotes.

Table 3 summarizes some of the important evolutionary diversifications from 3.4 billion years ago through to 400 million years ago in relation to the development of an oxidizing atmosphere. Clearly the early part of botanical history was aquatic, to be followed by an explosion of terrestrial diversification about 400 million years ago. Most of the aquatic forms were, and still are algae and Cyanobacteria. Most terrestrial forms are vascular plants, bryophytes and fungi. Hence the subsequent treatment of aquatic and terrestrial modes of life will be concerned with distinctly different taxa.

Table 3. Important historical events leading to the colonization of land.

Years Ago	*Event*	*Atmosphere*
4.7-4.8 billion	Origin of life Strong UV radiation leading to chemical evolution	Reducing atmosphere from volcanic activity
3.5-3.8 billion	Diversification of Prokaryota Autotrophy and heterotrophy present	
1.8 billion	Blue-green algal photosynthesis produces O_2 which is partially precipitated as ferric oxide, also liberated from water	Oxygen escapes from water into atmosphere
600-650 million	Diversification of aquatic eukaryotes. Lethal UV screened from surface waters	Oxygen accumulation continues
395-405 million	First land plants Lethal UV radiation screened from land	Oxygen at 1-10% of modern atmospheric level

2

The Aquatic Environment

There was life in water for perhaps two billion years before land was colonized. The antiquity of life in water is reflected in the great diversity of ancient higher taxa (divisions) in the sea. In contrast, the terrestrial environment is dominated by relatively few divisions and these are of recent origin. In this chapter consideration will be given to the physical and chemical properties of water as they relate to living organisms. In addition, the major ecological assemblages of living organisms will be introduced.

Physico-chemical Climate

Water is so common in everyday life that most people do not appreciate its unique properties. To begin with, water is liquid rather than frozen or gaseous over most of its occurrence on earth. Liquids are very rare in nature. Apart from water, only mercury occurs as an inorganic liquid on the earth's surface. Liquid water is the major component in the mass of all living organisms. Liquid water will dissolve more substances than any other solvent. This is important for living organisms because substances in solution provide the means for sustaining life. Oxygen and essential minerals are supplied in solution. Furthermore, toxic wastes are disposed of in solution.

The thermal properties of water are also essential to life. Water has a high **specific heat**. This means that large amounts of heat are required to raise the temperature of water in comparison with, for example, air. Only ammonia, liquid helium and lithium have higher specific heats. This means that aquatic environments fluctuate in temperature (through seasonal heating or cooling) to only a small degree. The sea is thus an equable environment and most marine organisms have no temperature regulating systems.

Water also has a high **latent heat**—more than any other substance at ordinary temperatures. This means that changing liquid water to steam or ice requires the addition or extraction of large quantities of heat. This characteristic bestows stability on bodies of water and the organisms that inhabit them. In addition, the density charactersitics of water ensure that freezing is restricted to shallow surface layers. This is because water has its greatest density at 4°C. When frozen, water floats on a warmer, denser liquid layer. This means that, even in the High Arctic, there is open water beneath a layer of less dense ice.

Water density depends on salinity and pressure, as well as temperature. A striking feature of aquatic systems is density stratification where water masses of different densities become layered, one above the other. A primary cause of stratification is heating of surface water layers in the spring so that a less dense region of surface water essentially floats on a denser, colder region. Often there is a sharp discontinuity between the two layers of differing water density (and temperature). This discontinuity is called a **thermocline**.

Vertical stratification of water columns is of critical importance to the life that they contain. The warm surface layer of water which forms in the summer time is essentially isolated from deeper waters. Many of the minerals that are essential for the growth of plants become incorporated in organic materials which fall through the thermocline. In the deeper waters organic remains are mineralized by microorganisms, but the ions required for plant growth are trapped in the cold, dark regions for much of the summer.

Since vertical stability or stratification has such important effects on the life of marine plants, destabilizing forces will also be important. Vertical mixing due to turbulence replenishes the minerals in the upper water layers. There are three main causes of vertical mixing: (a) orbital motion of waves at the surface, (b) convection currents caused by cooling and evaporation at the surface, and (c) currents passing over an uneven sea floor.

Horizontal transport of water is also important to plant growth. The complexity of ocean currents is beyond the scope of this book, but some mention of horizontal flow is essential. Currents are moved by wind, temperature differentials, evaporation and precipitation. Friction and the earth's rotation influence direction. Ocean currents cause upwelling of nutrient rich water in certain areas of the world. Along the equator in the Pacific Ocean the equatorial currents move in opposite directions causing a divergence and the upwelling of deep water. Water can also be displaced by offshore winds which cause localized upwellings of deep water (Fig. 3). Vertical mixing, whatever its origins, is a mixed blessing to plants suspended in the water column. Surface water containing plant cells is often moved downward. Since light is attenuated rapidly in water, these plant cells will soon become light deficient and stop growing. The relationship between photosynthesis, light and depth is shown in Fig. 4. In shallow water, which is relatively well illuminated, photosynthetic production of carbohydrate exceeds respiratory consumption, and growth can proceed. At the **compensation** depth (Fig. 4) photosynthetic production and respiratory consumption are equal. Below this depth respiration exceeds photosynthesis and growth must soon come to a stop. Water absorbs light much more rapidly than the atmosphere. Through its whole passage through the atmosphere solar radiation is reduced by only about 50%. In the clearest ocean waters attenuation of surface light by 50% occurs at less than 10 meters depth. This means that photosynthetic production is restricted to shallow waters, which, because of vertical stratification, become deficient in nutrient supply.

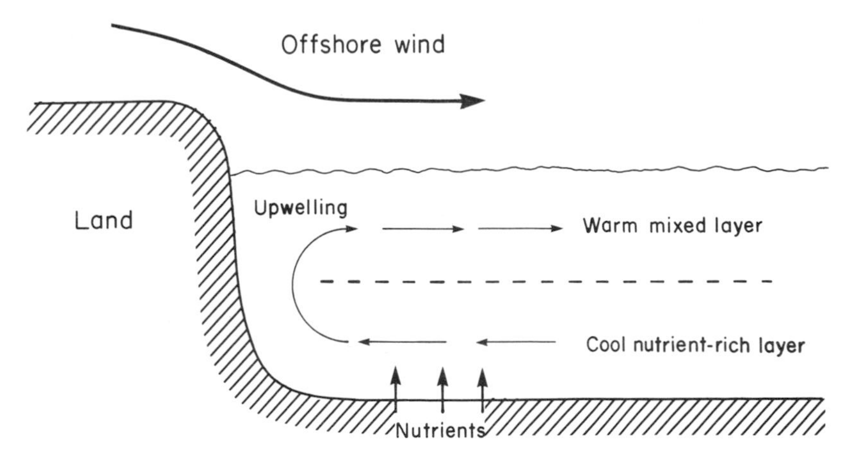

Figure 3. Upwelling of nutrient-rich water caused by offshore winds (modified after Mann, 1982).

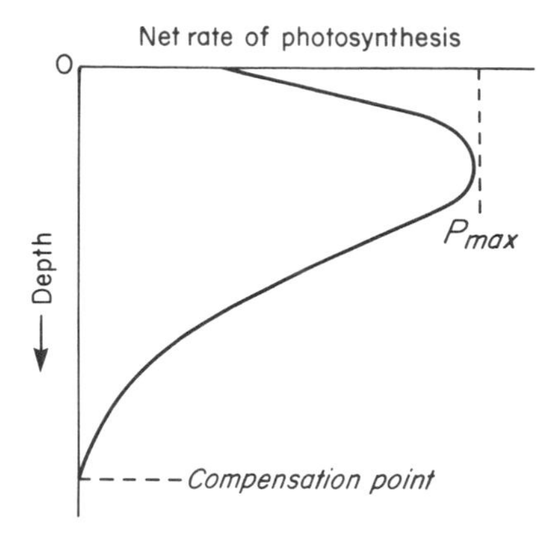

Figure 4. The relationship between phytoplankton photosynthesis and water depth. In shallow water photosynthesis is inhibited by high light levels (modified after Mann, 1982).

The nutrients which may limit the growth of aquatic plants are part of the complex chemistry of natural waters. Among the inorganic components, those which most affect productivity are nitrogen and phosphorus. Nitrogen gas in solution is available to only a few prokaryotes. All eukaryotes use combined nitrogen, most commonly as NH_4^+, NO_2^- or NO_3^-. Only prokaryotes have the enzyme system required for fixing molecular nitrogen. Among the inorganic ions, silicate, iron, manganese, cobalt and magnesium are also biologically important in that they may be scarce spatially and/or temporally. In the same way, certain organic components in solution affect the growth of plants. Many algae have vitamin requirements for growth (e.g., vitamin B_{12}).

Plants require more elements or compounds than those listed above. However, the supply in aquatic systems of many of these nutrients exceeds the demands for plant growth. For example, the amount of CO_2 (required for photosynthesis) is very large in comparison with other atmospheric gases. This results from the complex equilibrium that exists among various forms of inorganic carbon in water:

$$CO_2 + H_2O \leftrightarrow \underset{\text{carbonic acid}}{H_2CO_3} \leftrightarrow H^+ + \underset{\text{biocarbonate}}{HCO_3^-} \leftrightarrow \underset{\text{carbonate}}{2H^{++} + CO_3^{2-}}$$

Most of the CO_2 entering water becomes converted to the bicarbonate ion which is used directly in photosynthesis.

The discussion of the physico-chemical climate in the sea to this point has revolved around conditions in open water where plants live in suspension. Around the fringes of the oceans there exists another environment. The intertidal zone and the shallow subtidal support the growth of seaweeds. Hydrodynamic conditions at the ocean fringes produce an environment for plants which is quite different from that in open water. Seaweeds occupy one or more of the regions of water motion shown in Fig. 5. In the uppermost water layer, the current zone, water movement is unidirectional at about 1 m/s. Below the current zone is the surge zone where water is oscillating in a motion produced by waves. In the lowermost 2 cm above the sea floor is the boundary layer where water movement is slowed by friction to $<10\%$ of current and surge velocities.

Classification of Plant Habitats

As indicated above, plants in aquatic environments are either in suspension or attached to the sea (or lake) floor. This is the most straightforward classification of aquatic plant habitats—plants in suspension are **planktonic**, those which are attached are **benthic**. The benthic environment may be conveniently subdivided into soft and hard rock substrata. In the sea the hard rock substrata are often dominated by seaweeds down to the compensation depth for photosynthesis. Soft bottoms in the illuminated zone are dominated by vascular plants and microscopic algae.

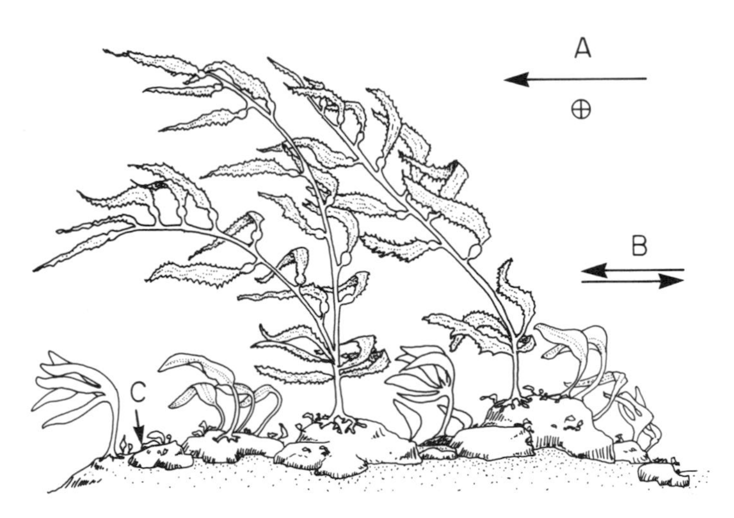

Figure 5. Water motion regions in a California kelp forest. A. Current zone; B. Surge zone; C. Boundary layer (modified after Neushul, 1972).

The plankton is dominated by microscopic algae. In nearshore environments many of these plants spend at least part of their lives in the benthos (on the bottom) and are called **mero**planktonic. In open ocean water the planktonic algae do not have benthic stages and are called **holo**planktonic. It should be pointed out that the distinctions among all of the aquatic lifestyles of plants are artificial and tend to intergrade. Thus many planktonic species may be found in the benthos. Similarly, most benthic seaweeds have planktonic reproductive stages. Bearing in mind this qualification, the botanical diversity of aquatic habitats will be introduced in the two following chapters as phytoplankton and seaweeds.

3

Phytoplankton in the Sea

The vegetation of most places on land is fixed to the ground. In the sea most plant life consists of tiny organisms that are not fixed, but moving in and with the water masses. The size of phytoplankton and the oceanic world in which they live places them outside normal human experience.

Life at Low Reynolds Number

Most planktonic plants are very small. The smallest are called **nanoplankton** and are in the size range of 2-20μm. These plants are not much bigger than bacteria, and they may form the bulk of phytoplankton biomass in large tracts of the world's oceans. **Microplankton** are in the size range 20-200μm and **macroplankton** are in the range of 200-2000μm. Even the largest phytoplankton plants are only in the 2 mm size range and these forms are relatively rare. Most marine plants are nanoplankton and microplankton. The physical consequences of being so small in a fluid medium are very great. In relation to that of larger organisms, the world of the phytoplankton is a very viscous place—a world of low Reynolds number. Reynolds number is the ratio of inertial forces/viscous forces. For a human swimming in water the Reynolds number is about 10^4. For a goldfish the number is 10^2. For even quite large phytoplankton the number is only about 10^{-2}. What this means is that inertia is irrelevant for small organisms and viscous forces are dominant. Living in water at such a low Reynolds number is like swimming in molasses.

Functional Diversity of Phytoplankton Morphologies

There is an enormous diversity in the shapes and sizes of marine phytoplankton (Fig. 6). It is only in the last 20 years that a real understanding of the functional significance of some of this diversity has emerged. The shapes and sizes of the organisms appear to be, in part, adaptations to (a) movement in the water column and (b) to grazing pressure. Because phytoplankters are photosynthetic, they need to stay in the illuminated zone, but because nutrients are critically limiting for growth, the plants need to move through the water. Shapes and sizes are related to these two needs and not simply to flotation. If all cells were to float in the well illuminated surface layer of water, all nutrients would be quickly exhausted, and growth would stop.

The phytoplankters shown in Fig. 6. vary morphologically in the following ways: (a) size, (b) presence or absence of flagella, (c) solitary or joined in colonies, (d) presence or absence of cell protuberances, and (e) flattened or rounded. The organisms also differ cytologically and physiologically in ways that affect their buoyancy properties.

Some of these morphological features may be related to the gradient of turbulence/nutrient conditions in the sea. Increased turbulent movement of water is associated with increased nutrient supply (Fig. 7). At the lower end of the linear gradient (low nutrients and low turbulence) the plants are usually large and motile (flagellate), like *Ceratium* in Fig. 6. Under high nutrient, high turbulence conditions there is a predominance of small, non-motile cells, like *Chaetoceros* (Fig. 6). In what ways is *Ceratium* adapted to low nutrient/low turbulence conditions? First of all, the cells are more dense than water and will therefore sink in water that moves very little. Calculations and direct observations show that large cells of 100-200μm diameter sink about 1 m/day in still water. Large organisms sink faster than small ones, so the sinking problem is magnified for the large celled species like *Ceratium* that predominate in low turbulence water. In fact *Ceratium* can swim, using its flagella, at 6-20 m/day and can easily counteract gravitational sedimentation. Large celled flagellates can even exhibit daily vertical migration patterns, illustrating the independence of these from the forces of gravitational sinking. *Ceratium* and several other genera

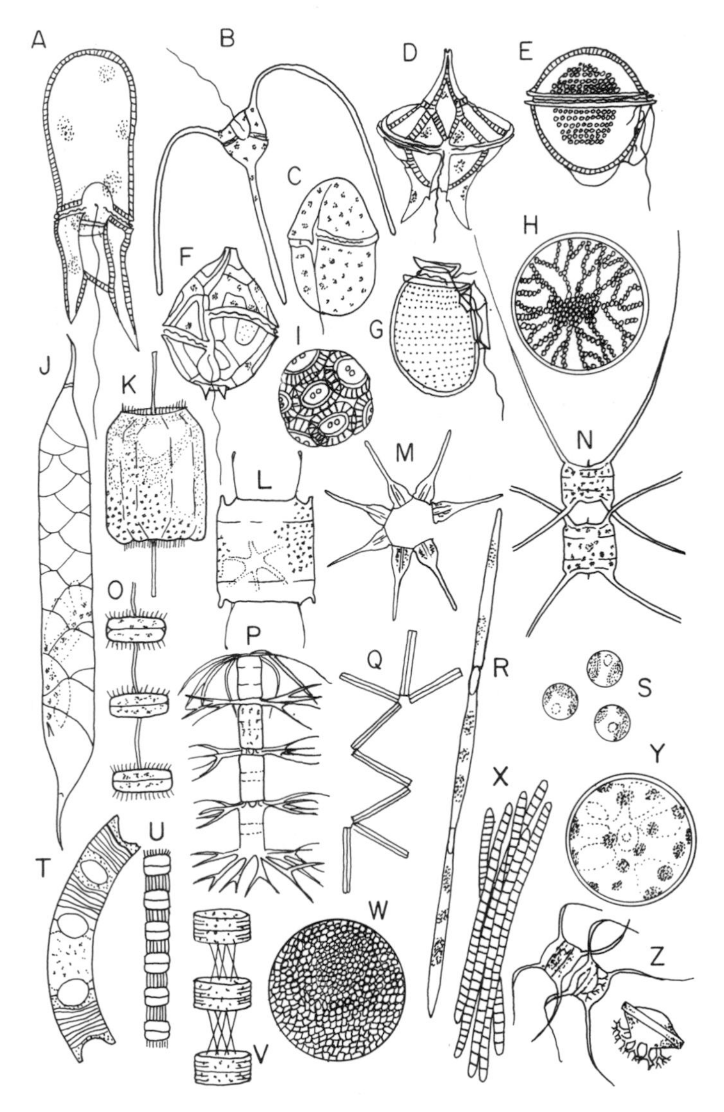

Figure 6. Common microscopic genera of marine phytoplankton. A., B. *Ceratium;* C. *Gymnodinium;* D. *Peridinium;* E. *Phalacroma;* F. *Gonyaulax;* G. *Dinophysis;* H. *Pyrocystis;* I. *Coccolithus;* J. *Rhizoselenia;* K. *Ditylum;* L. *Biddulphia;* M. *Asterionella;* N. *Chaetoceros;* O. *Thalassiosira;* P. *Bacteriastrum;* Q. *Thalassiothrix;* R. *Nitzschia;* S. *Phaeocystis;* T. *Eucampia;* U. *Skeletonema;* V. *Coscinosira;* W. *Coscinodiscus;* X. *Trichodesmium;* Y. *Halosphaera;* Z. *Chaetoceros* (modified after Round, 1981).

shown in Fig. 6 (e.g., *Peridinium, Gymnodinium*) have a free flagellum used for swimming and another flagellum wrapped around their peripheries like loose robe cords. This second flagellum (called the transverse flagellum) is larger and does more work than the other flagellum (used for swimming). The transverse flagellum causes the cell to rotate and the flattening of the cell increases the resistance to rotation. The net effect is an increase in turbulence and water exchange at the cell surface. The activities of the two flagellum types in cells such as these seem crucially related to obtaining dissolved nutrients from mineral depleted waters, rather than to staying in suspension.

At the high turbulence/high nutrient end of the marine spectrum (Fig. 7) the phytoplankton is dominated by a group of organisms called diatoms (in the division Chrysophyta in Table 2). *Chaetoceros* is a diatom (Fig. 6). Apart from their flagellate sperms, members of this group of organisms are unable to swim through water. Furthermore, their cell walls (called frustules) are very heavy because of the silica (glass) content. These two features ensure that they will sediment out rapidly in still water, but in turbulent waters of the sea, they stay in suspension.

When the wind blows across the surface of the water at velocities exceeding 3m/s convection cells are created in the water. In cross-

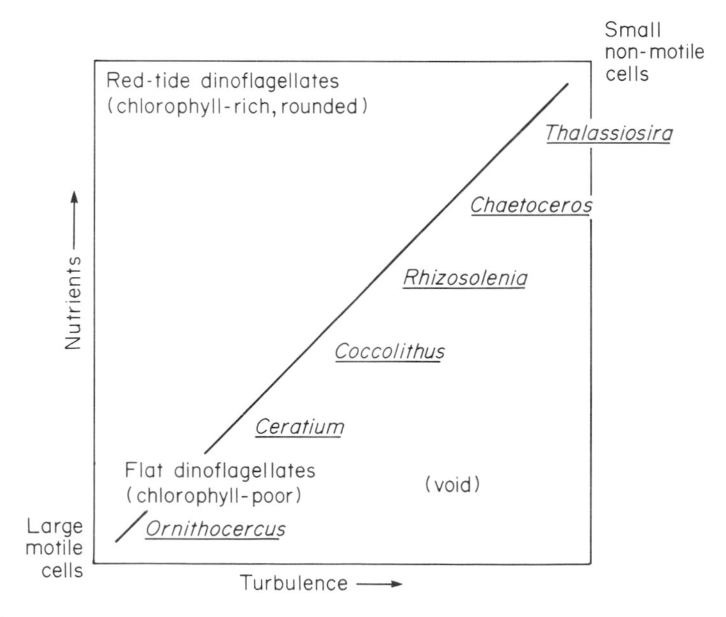

Figure 7. Gradient of turbulence and nutrient concentrations in the sea, and planktonic cell types found along the gradient (modified after Margalef, 1978).

section the convection cells are seen to occur in oppositely rotating pairs (Fig. 8). The phytoplankton cells become entrained within the rotating convection cells. Entrainment occurs, even when plant cells are denser than water, provided that the maximum upward water velocity is greater than the sinking velocity of the plant . In fact, at windspeeds of only 3-6 m/s, the upward water velocities are 2-3 orders of magnitude greater than the cell sinking velocities. This means that heavy, non-motile diatoms are kept in suspension through water movement. Under turbulent conditions phytoplankton cells of all shapes, sizes and densities are readily entrained in the convection cells near the surface.

This brings us to the next point for consideration. Why are non-motile phytoplankters so well equipped with spines and other cellular appendages? It has been known for a long time that appendages slow sinking rates, and the structures came to be seen as mechanisms that helped cells to stay in suspension. However, the best developed appendages occur in species that live in turbulent waters where biological suspension mechanisms are irrelevent. Cellular appendages are best viewed as devices that ensure tumbling, twisting and rotational movement of cells through water. This type of movement will break up micronutrient gradients that exist around each cell. Phytoplankters drain the nutrients from the microzones that surround each cell. Replacement by diffusion is slow and any movement of the cell will enhance renewal of the surrounding water. The projecting spines of *Chaetoceros* (Fig. 6), while tending to orient the cells with their long axes perpendicular to the plane of descent, also induce slow rotational movements.

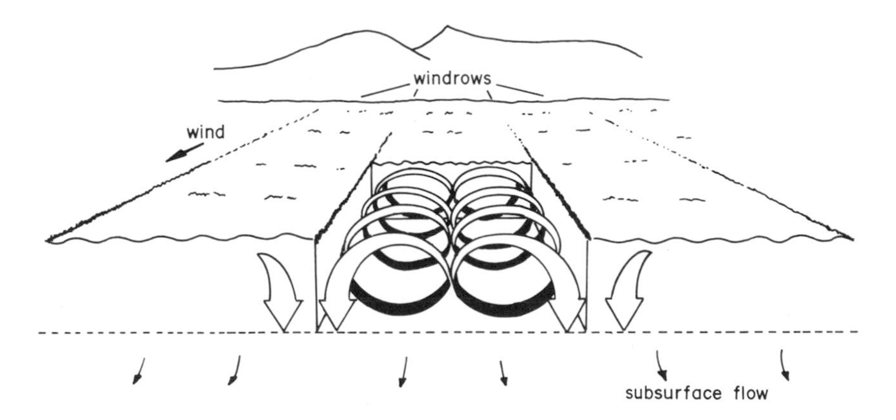

Figure 8. Orbital rotation of wind driven surface water movement (modified after Reynolds, 1984).

Many non-motile phytoplankters form chains of cells , and again, this appears to facilitate rotation of the entire colony in a moving medium. In addition, chains of cells usually sink faster than unicells. The rotating, rapid sinking of cell chains may provide a forced convection of nutrients and take the plants into local patches of nutrient rich water. Provided that the sinking rate is less than the upwelling water velocity, any downward movement of cells through sinking will enhance nutrient supply with no risk of sedimentation out of the illuminated zone of the sea.

From the account given above, it is clear that much of the morphological diversity of of phytoplankton (shown in part in Fig. 6) might be related to the problems of obtaining nutrients in the photosynthetic zone of the sea. Another, entirely different selective pressure is grazing or herbivory. Feeding preference experiments with zooplankton appear to indicate that some of the morphological features of phytoplankters might be anti-grazing devices. Cell appendages and colony formation may be grazing deterrents. However, this is not a universally accepted point of view. One morphological feature of plankton that does appear to be related to herbivory is the presence of a mucilage sheath around the cell periphery. It has been shown that many species of ensheathed green algae (Chlorophyta) readily withstand passage through the guts of grazing crustacean zooplankton. Experiments have shown that this group of green algae increased in proportion to other kinds of phytoplankters in the presence of grazers. Large species (>40 μm) appear unaffected by grazing, while small species are suppressed. It should be pointed out that these results were obtained in a small (17 ha.) lake. In the sea, other types of grazers (e.g. planktivorous fish) might change the relationships between susceptibility to grazing and phytoplankton morphology.

Cytological and Physiological Adaptations to Life in Suspension

To this point discussion in the chapter has centered on the diversity of various cell and colony morphologies. There are also cytological and physiological features that are clearly related to the planktonic mode of existence. Prominent among these is the occurrence of **gas vacuoles** in the cells of blue-green algae (Cyanobacteria). Among marine blue-green algae the vacuoles are found in the filamentous form *Trichodesmium* (Fig. 6) which is very common in the tropics. This species grows very slowly and must remain in suspension. Gas vacuoles ensure that it does. These vacuoles are minute, gas filled, membrane enclosed structures occurring within the cytoplasm. If the vacuoles are collapsed by a pulse of hydrostatic pressure, the filaments will sink. Freshwater planktonic species with gas vacuoles can regulate their buoyancy through forming or collapsing vacuoles. However, the marine genus *Trichodesmium* is unable to do this. The membranes of the vacuoles of this organism are two to six times stronger than those of freshwater forms. The stronger vacuole membrane may be an adaptation that prevents hydrostatic pressure from collapsing vacuoles during the deep mixing (up to 100m) that may occur in ocean storms.

Recent work has shown that some marine species of phytoplankton may regulate their buoyancy through ionic exchange. The mechanism does not work in freshwater. *Ditylum*, a marine diatom, appears to selectively absorb monovalent ions that reduce the cell density to less than that of seawater, resulting in positive buoyancy.

The role of flagella in the movement of phytoplankters has already been mentioned. However, the way in which they work has not been described. Flagella are really intracellular structures. That is, they are enclosed within the envelope of the cell membrane. A longitudinal section through a flagellum is shown in Fig. 9. Notice the tubular structures running the length of the flagellum. These are the microtubules. In cross-section (Fig. 9) it can be seen that there are eleven tubular sets, nine around the periphery (these are doublets) and two in the center. This type of construction is called **9 + 2**, for obvious reasons. The basis for movement is an active sliding (up and down) between the peripheral doublet microtubules which, when resisted by structures in the flagellum, cause bending. The central tubules are not involved in sliding or bending. Recent evidence shows that in the tiny flagellate

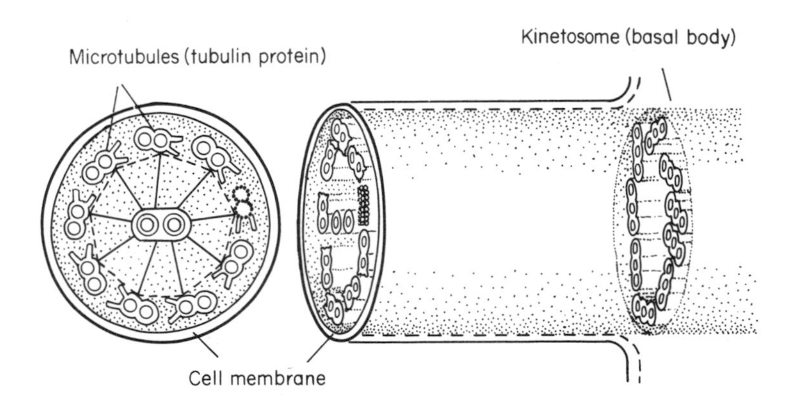

Figure 9. Construction of the eukaryote flagellum (modified after Margulis, 1982).

Micromonas the central tubules rotate continuously in a counterclockwise direction. The cause of rotation is unknown, but there is speculation about a rotary motor near the flagellum base. The spinning central tubules appear to regulate flagellum motion. Mutant flagella without central tubules, but with a full complement of peripheral tubules, are immobile. It has been proposed that the systematic movement of the central tubules is involved in regulating flagellar motion, possibly by determining which peripheral doublets undergo active sliding at any particular phase in the beat cycle.

Whatever the details of microtubule sliding are, flagella move the cells to which they are attached by bending. Cells have helical propeller drive or whiplash drive (Fig. 10). *Micromonas* has a helical drive system. This was shown experimentally when cells were stuck to glass slides by their flagellar tips. The cell bodies were seen to rotate around the fixed flagellum. With the left handed twist of the flagellum, the rotary movement ensures propeller drive. The whiplash flagellum is really a type of flexible oar. Rigid oars will not work at low Reynolds number because they move in a reciprocal fashion. In a sticky medium with low inertial forces any structure moved in a reciprocal manner exactly retraces its trajectory back to where it started and there is no net forward propulsion. But with a flexible oar, the structure bends one way during the first half of the stroke and the other way during the second half. Any kind of asymmetric motion will ensure that swimming occurs.

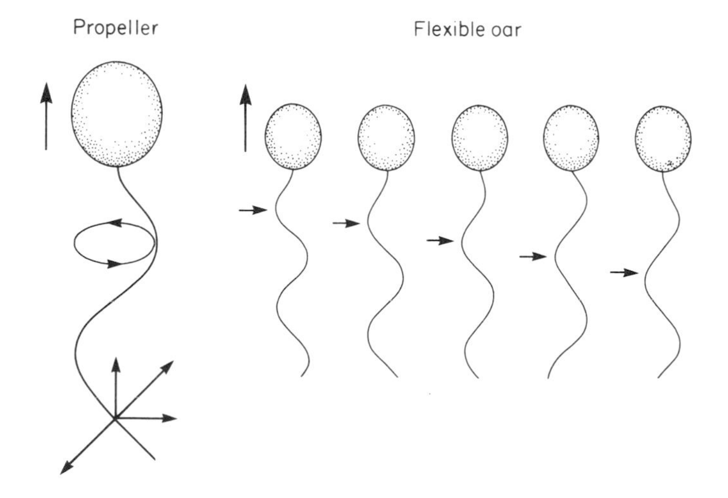

Figure 10. Propeller and flexible oar drive by flagellar action (modified after Chapman, 1979).

Minerals That Limit Phytoplankton Growth

There is something of a consensus opinion that many of the morphological and cytological features of phytoplankton once thought of as flotation devices are, in fact, mechanisms than ensure movement through nutrient deficient waters, thereby increasing nutrient supply. What are the marine nutrients that limit plant growth? In most locations examined, nitrogen is the first nutrient depleted during phytoplankton growth. Nitrogen is an absolute requirement for the formation of the protein and nucleic acid components of all cells. In the eastern South Pacific ocean, silicic acid may be depleted before available nitrogen. Phosphate may the most limiting nutrient for plant growth in the Mediterranean. Phosphorus is an absolute requirement for all living cells, but silicic acid is a special nutrient for diatoms which have silica as a major component of their cell walls. Silicic acid is rarely required among other groups of algae.

Nitrogen is available in many forms in the sea. Nitrogen gas in physical solution is utilized only by nitrogen-fixing prokaryotes. This supply of N is virtually unlimited for blue-green algae that have an N-fixing capability. Eukaryotic organisms are much less fortunate. The major sources of N that they are able to use are NH_4^+, NO_2^-, NO_3^- and the organic compound urea. Other sources of N cannot be utilized readily by these organisms and their growth is often severely limited in the sea by ambient nutrient concentrations. There is, however, a recent hypothesis which proposes that phytoplankton cells may be N sufficient in waters of low nutrient concentration. According to this hypothesis, the cells obtain their nutrients from microzones (of microliter volumes) of elevated nutrient concentration that cannot be detected with modern instrumentation. The problem with hypotheses such as these is that there is very little information available on the growth of phytoplankters in the sea. Nearly all growth rates are derived from photosynthetic rates measured in closed chambers. However, photosynthetic rates and growth rates are often uncoupled in time. Furthermore, closed chamber experiments hardly simulate the convection cell currents of the open ocean.

An alternative approach to the use of growth rates as indicators of nutrient limitation lies in the measurement of cell chemical composition. The internal cellular concentration of a nutrient has been shown to have a hyperbolic relationship with growth rate in culture. If this relationship is known, then in theory, measurements of cell concentrations of nutrients in the wild will correlate with growth rates in the sea.

Unfortunately very few species of oceanic phytoplankton can be grown in culture, so unequivocal data are unavailable.

Nitrogen fixation in blue-green algae is usually carried out in specially differentiated cells called heterocysts (Fig. 12). These cells occur in filaments and are recognizable by their enlarged size and pale color. The cells also have thickened walls. Nitrogen gas is absorbed by cells adjacent to the heterocyst and is transferred in solution to the heterocyst interior. Nitrogen is then reduced to NH_3 in a process requiring energy (ATP is hydrolyzed) and catalyzed by nitrogenase (an enzyme). The process will not operate in the presence of oxygen and it is thought that the heterocyst walls act as oxygen barriers. The photosynthetic machinery of heterocysts is incomplete so that oxygen is not released. However, the photosynthetic pigments that are present produce ATP on absorption of light, and this might energize the N_2 reduction process. Photosynthetic CO_2 reduction does not occur in heterocysts.

Unlike nitrogen, phosphorus and silica are each available to phyto plankton in only one main form. Phosphorus is available as PO_4^{3-} and silica as $Si(OH)_4$. Phosphorus is accumulated in cells in a storage form for later use. Nitrogen is also accumulated (in various forms) in phytoplankton cells. Silica is not stored, and uptake is accelerated when cell division and new wall formation occur in diatoms.

There is a huge literature on the ways in which limiting nutrient concentrations affect uptake velocities, growth rates and storage rates in phytoplankton. Most of this information is beyond the scope of this book. Surprisingly little information is available on the mechanisms of nutrient uptake at the cell surface. Current opinion favors a proton-

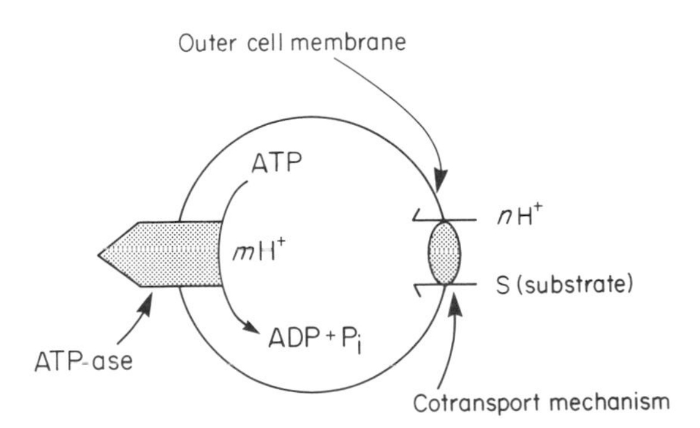

Figure 11. Model for uptake of substrate (S) by proton-linked cotransport through cell membrane. Driving force is a proton gradient across the cell membrane. ATP-ase (enzyme) pumps out protons (mH$^+$ per ATP molecule hydrolyzed). Transport of nutrient molecule (S) is linked to cotransport of n protons (H$^+$) (modified after Syrrett, 1981, by permission of the Minister of Supply and Services, Canada).

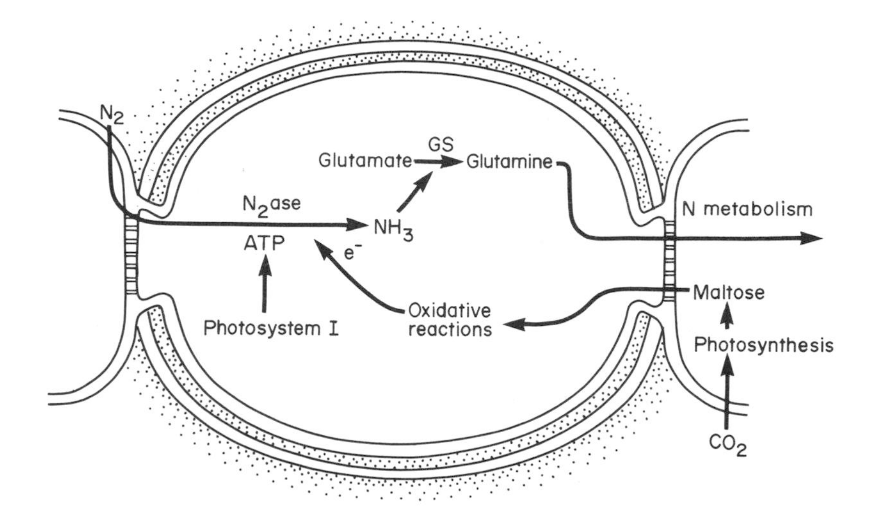

Figure 12. Nitrogen and carbon assimilation associated with nitrogen fixation in heterocysts. N_2ase = nitrogenase; GS = glutamine synthetase; photosystem I = part of photosynthetic machinery; e^- shows electron transport (modified after Haselkorn, 1978. Reproduced, with permission from the Annual Review of Plant Physiology Vol. 29. ©1978 by *Annual Reviews; Inc.*).

linked cotransport (a model is shown in Fig. 11). According to this idea the driving force for nutrient uptake is a proton gradient across the cell boundary which is maintained by cell metabolism. On the left of Fig. 11 is an enzyme ATP-ase whose function is to pump out protons (mH$^+$ per ATP molecule hydrolyzed). On the right is a cotransport system in which the transport of one nutrient molecule is linked to the cotransport of n protons. The driving force is (a) the charge difference across the membrane and (b) the pH gradient. ATP maintains these gradients across the membrane. The ATP is derived from the photosynthetic mechanism, therefore light is required for nutrient uptake. For reasons explained above, nutrients in the sea are in the lowest concentrations in the illuminated zone. This unfortunate combination of conditions certainly complicates the lives of phytoplankton.

Systematic Diversity of Phytoplankton

The morphological diversity of phytoplankton cuts across taxonomic boundaries. Only blue-green algae (Cyanobacteria) are distinguished by unique functional adaptations (gas vacuoles and heterocysts) related to the hydrodynamic and nutrient climate of the sea. Flagella are found in most phytoplankton groups, and so are many of the functionally important characteristics described above. However, the major groups (divisions) of marine algae have very different evolutionary histories. The phylogenetic diversity has resulted in a huge taxonomic diversity in the upper hierarchical categories. This taxonomic diversity is classified on the basis of biochemical and taxonomic characteristics shown in Table 4.

KINGDOM PROKARYOTA

Division Cyanobacteria

Blue-green algae (Cyanobacteria) have a special place in the evolutionary history of the planet. These organisms were the first to produce oxygen using a process that led to the transformation of the earth's atmosphere from reducing to oxidizing conditions. Water is the hydrogen donor (and oxygen source) in the photosynthesis of blue-

Table 4. Diagnostic features of algal divisions treated in Chapters 3 and 4 as phytoplankton and seaweeds.

Division	*Plankton or Benthos*	*Major Accessory Pigment**	*Storage Product*	*Flagellation*
Cyanobacteria	Mostly plankton	Phycobilins	Myxophycean starch	None
Rhodophyta	Mostly benthic	Phycobilins	Floridean starch	None
Chlorophyta	Both	Chlorophyll *b*	Starch	2,4 or more anterior, smooth†
Euglenophyta	Plankton	Chlorophyll *b*	Paramylon	2, one or both emergent with hairs
Pyrrophyta	Plankton	Chlorophyll *c* peridinin	Starch	2, equal length without hairs
Chrysophyta	Plankton	Chlorophyll *c* fucoxanthin	Chryso-laminaran	1-3, anterior with hairs
Phaeophyta	Benthos	Chlorophyll *c* fucoxanthin	Laminaran	2, lateral, smooth, with hairs
Cryptophyta	Plankton	Chlorophyll *c* phycobilins	Starch	2, lateral with hairs

* Accessory pigments absorb light and transfer energy to chlorophyll *a*.
† Algal flagella either have lateral hairs or they may be smooth.

green algae. The cell organization of blue-green algae is prokaryotic (Fig. 13). Around the periphery of the cell is a complex **peptidoglycan** wall made up of amino acids and sugars. The wall chemistry is more complex than those of eukaryotic algae.

Within the cell the most prominent feature is the collection of photosynthetic membranes called **thylakoids**. These are flat sacs which contain the fat soluble photosynthetic pigments (chlorophyll *a* and several carotenoids). The water soluble photosynthetic pigments (phycobilins) are contained in small packets called **phycobilisomes** attached to the outside of the thylakoid membranes. Light energy trapped by the phycobilin pigments is transferred to chlorophyll *a* which is the primary photochemical pigment in a process that leads to the production of ATP and reducing power (in the form of NADPH). The products of the photochemical reaction are used in the cytoplasm of blue-green algae to reduce CO_2 to sugars.

Apart from the thylakoids, the other prominent features of blue-green algal cells are **gas vacuoles, cyanophycin granules, polyphosphate bodies, nucleic acid fibrils, glycogen granules** and **polyhedral bodies.** The role of gas vacuoles has already been described. Cyanophycin granules are nitrogen storage bodies (co-polymers of aspartate and arginine) that are particularly noticeable in nitrogen fixing species. Polyphosphate bodies are phosphorus storage granules. Nucleic acid fibrils contain the entire genetic complement of each cell. Glycogen granules form a

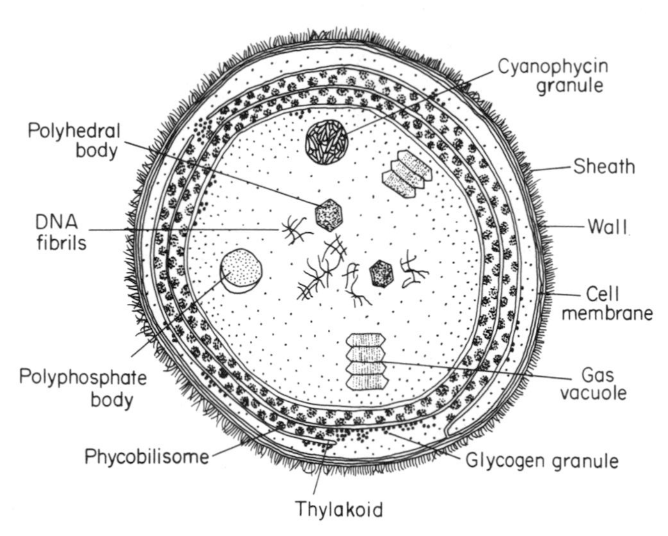

Figure 13. Fine structure of the blue-green algal cell (modified after Lee, 1980).

carbon reserve. Polyhedral bodies, also known as **carboxysomes**, are membrane enclosed structures which contain carboxylating enzymes concerned with the entry of CO_2 into the photosynthetic Calvin cycle.

There is very little morphological diversity among blue-green algae (Fig. 14). The organisms are unicells, colonies or filaments.

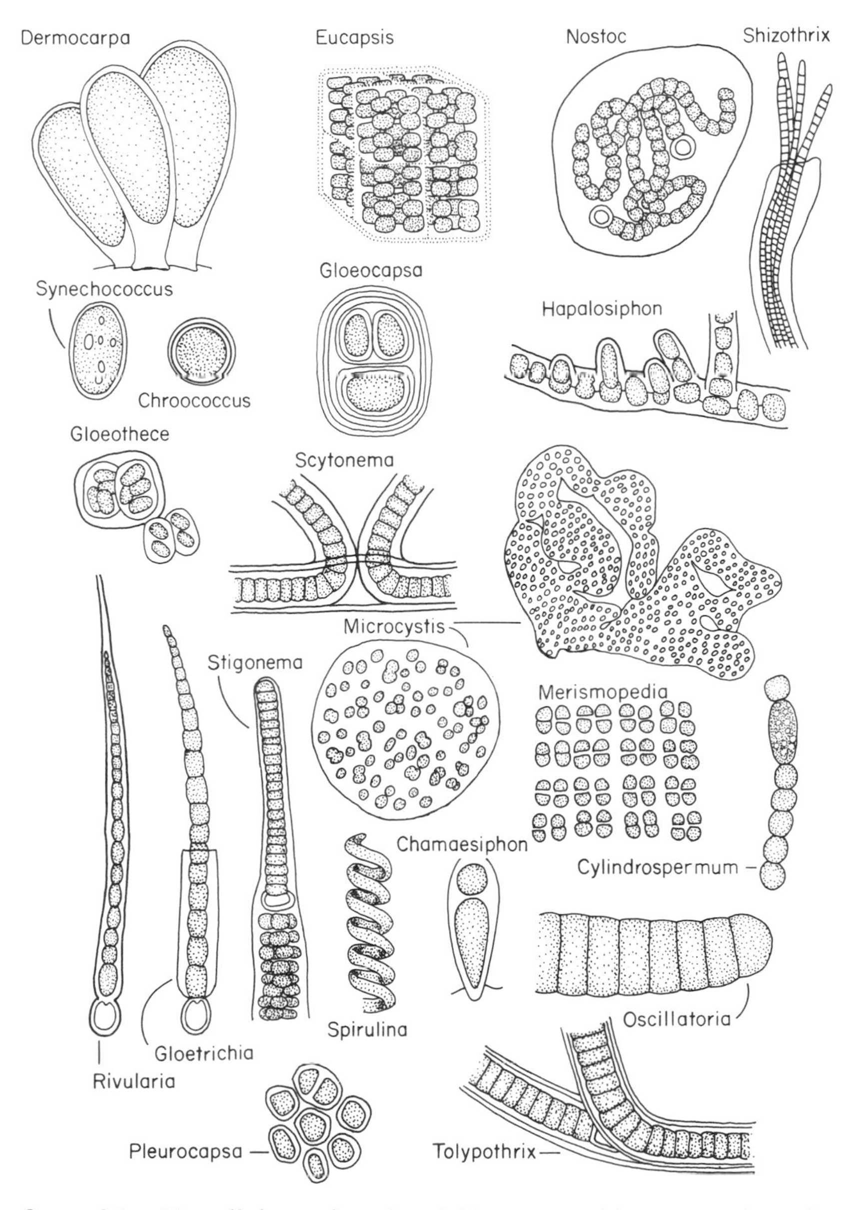

Figure 14. Unicellular, colonial and filamentous blue-green algae; h.c. = heterocyst (from *Plant Diversity: An Evolutionary Approach*. © 1965, 1969 by Wadsworth Publishing Co.).

Filaments are sometimes composed of morphologically identical cells. Differentiation may occur in the formation of heterocysts. The cells in filament forming species are interactive .

Reproduction in blue-green algae is asexual. In most unicells reproduction occurs by simple fission. In multicellular forms the chief method of reproduction is by fragmentation when the plant body breaks up and each component forms a new organism.

Differentiated reproductive structures are also formed. In some filamentous species the filaments differentiate into short sections that are released from the parent sheath. The short cell fragments have a gliding movement that is possibly powered by mucilage secretion. True spores are formed by intracellular differentiation in some species. The thin walled spores are released by breakdown of the parent cell wall, and each forms a new organism. True sexual reproduction does not occur in blue-green algae.

KINGDOM EUKARYOTA

Through about two-thirds of biological history only prokaryotic organisms existed on earth. The most ancient eukaryote groups are perhaps one billion years old. This is a very tentative figure because the fossil record offers so few clues. The way in which eukaryote cells evolved from prokaryote cells is matter for some speculation. According to the popular **endosymbiosis** theory, the organelles that characterize eukaryotic cells had ancestors that were free-living prokaryotes. The ancestral forms took up residence inside other prokaryotes. Endosymbiotic relations between modern eukaryotes and prokaryotes do exist, and this lends weight to the theory. According to the endosymbiosis theory the ancestors of the eukaryotic chloroplast were once free-living blue-green algae.

Division Rhodophyta

Most red algae are seaweeds. There are only about 10 genera that are unicellular and planktonic. Description of this division as a whole is reserved for the next chapter.

Porphyridium cruentum is a simple unicellular red alga (Fig. 15). Examination of Fig. 15 shows that this eukaryotic cell interior is compartmentalized into chloroplast, nucleus, mitochondria, golgi, etc., all of which are enclosed within unit membranes. Notice that flagella are absent here and throughout the members of this division. Within the chloroplast thylakoid, construction is quite similar to that of blue-green algae. Also embedded in the chloroplast is a **pyrenoid** which is a differentiated proteinaceous matrix that accumulates reserve products.

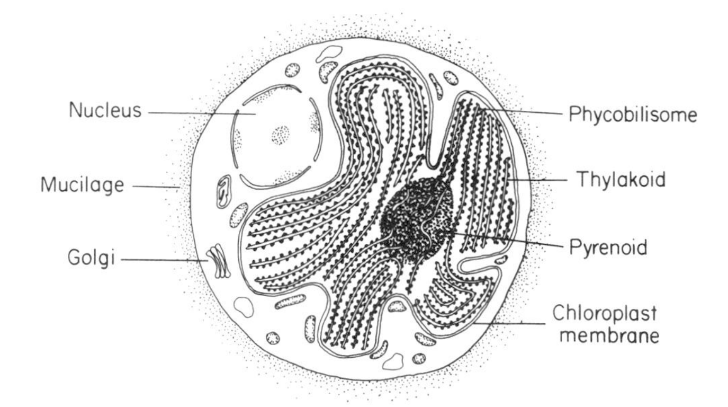

Figure 15. Fine structure of the unicellular red alga *Porphyridium* (modified after Lee, 1980).

The cell walls of red algae are generally composed of an amorphous matrix of galactans (including the commercially important agars and carrageenans) and a fibrillar cellulose component. In *Porphyridium* the fibrillar component is absent and the amorphous wall matrix is continually shed as a mucilage into the water.

Sexual reproduction is unknown in unicellular red algae and the organisms generally propagate by simple fission.

Division Chlorophyta

The Chlorophyta is viewed as the ancestral link between prokaryotic life in the Precambrian era (about one billion years ago) and the first vascular plants of 450 million years ago. Furthermore, the cell structures of modern green algae fall clearly into two groups. One group is most similar to vascular plants, and the other group is more similar to other algae. Members of the Chlorophyta are common in all aquatic habitats, especially those that are freshwater. Green algae are not as conspicuous in the marine phytoplankton as they are in freshwater. However, the marine seaweed flora is especially rich in green algae.

The cell structure of *Volvox*, a colonial green flagellate, is shown in Fig. 16. The features that distinguish the cells from those of red algae are the presence of a flagellum, an eyespot and thylakoids that are stacked (rather than single, as in red algae). The eyespot is part of the chloroplast, but it serves in the detection of direction of illumination. The cells swim toward or away from varying light intensities.

The morphologies of planktonic green algae (Fig. 17) range from unicells (motile and non-motile) to colonies (motile and non-motile). Benthic green algae have much more diverse morphologies, and these are discussed in the next chapter.

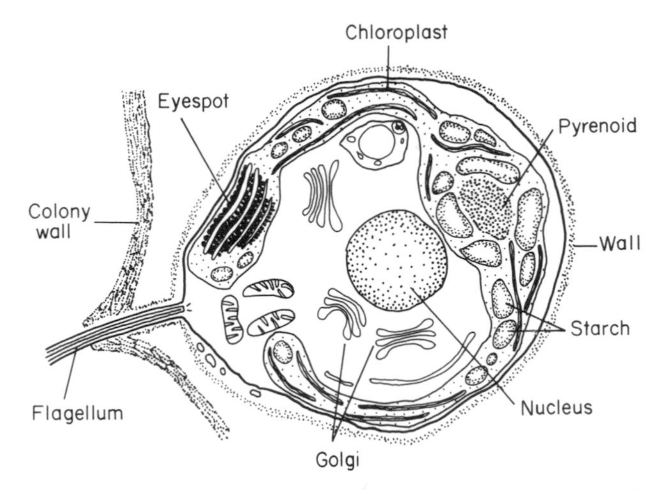

Figure 16. Fine structure of one cell in a *Volvox* (green alga) colony (modified after Lee, 1980).

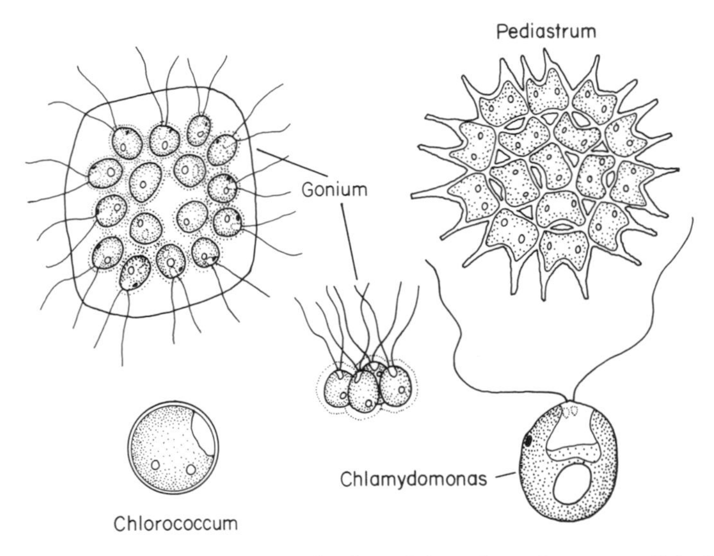

Figure 17. Some unicellular and colonial (motile and non-motile) planktonic green algae (from *Plant Diversity: An Evolutionary Approach.* © 1965, 1969 by Wadsworth Publishing Co.).

Asexual reproduction through flagellate spore formation is common throughout the Chlorophyta. Generally, vegetative cells round off their cytoplasmic contents into small spore units each equipped with flagella. The spores are released on disintegration of the parent cell wall. Asexual reproduction in colonial green algae is interesting. Complete minute colonies are formed within the parent cell walls. This is shown for *Volvox* in Fig. 18.

In the planktonic green algae sexual reproduction through fusion of gametes and subsequent meiosis is common. The gametes that fuse may be flagellate and identical in appearance. This is called **isogamy**. In **anisogamy** the female gamete is larger than the male, but both are flagellate. Some of the complex colonial forms have non-motile eggs that are fertilized by much smaller flagellate male gametes.

One important group of freshwater green phytoplankton, the **desmids**, does not have flagellate gametes. Instead, cellular fusion is achieved through conjugation tubes that form between parent cells, or through the release of non-motile ameboid gametes in a mucilage matrix. In most planktonic green algae meiosis occurs in the zygote so that vegetative cells are haploid.

Division Euglenophyta

Members of this division are microscopic flagellates. They have the same photosynthetic apparatus as green algae, but little else in common. One order (of three in the division) is entirely non-photosynthetic (the organisms obtain food by catching prey organisms, engulfing them through a permanent mouth, and digesting them internally). It is hard to avoid speculating on the evolutionary origins of the euglenophytes since they have so many animal and plant characteristics.

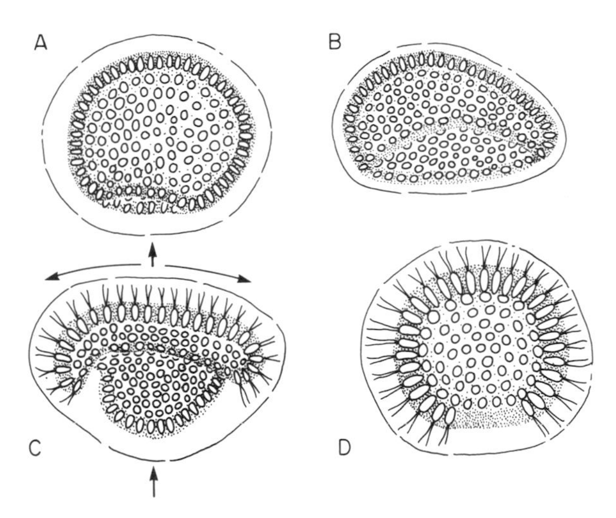

Figure 18. Development of a multicellular colony within a single mother cell in *Volvox* (green phytoplankter); A. Invagination beginning, B. Further invagination, C. Cells at upper end of daughter colony evaginated so that flagella point outward, D. Inversion near completion (from *An Evolutionary Survey of the Plant Kingdom.* © 1965 by Wadworth Publishing Co. Reprinted by permission of the publisher).

Although some euglenoids are photosynthetic, none of them has a cell wall. It seems entirely possible that euglenoids originated as heterotrophic flagellates in which chloroplasts are secondary inclusions.

The cells of euglenoids are bound only by the cell membrane (Fig. 19) which is ridged with proteinaceous bands to form a **pellicle**. The flagella emerge from a **reservoir** at the anterior end. The flagella bear hairs. In some forms two flagella emerge from the reservoir. In others, one flagellum is very short and not emergent. Many euglenoids have a large eyespot close to the reservoir, and adjacent to this is a crystalline paraflagellar body on the base of one of the flagella. The two structures are used in photonavigation. If the eyespot shades the **paraflagellar** body, the cell can determine the direction of illumination and adjust its swimming behavior.

Reproduction in euglenoids occurs by longitudinal cell division. Sexual reproduction is unknown.

Division Chrysophyta

This division contains three important phytoplankton groups. These are treated here as three separate classes.

Class Bacillariophyceae. Members of this class are called **diatoms**. The

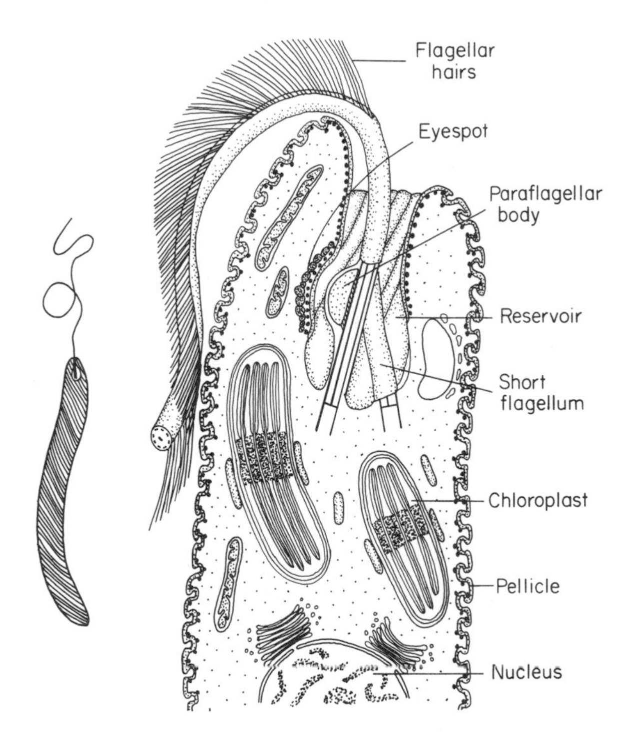

Figure 19. Fine structure of *Euglena* (modified after Lee, 1980).

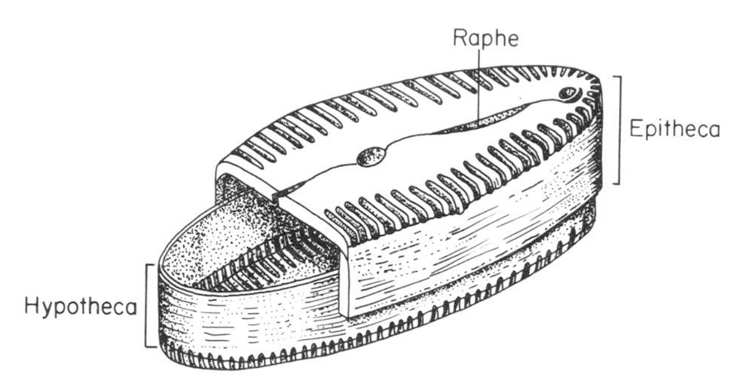

Figure 20. Frustule construction in a bilaterally symmetrical diatom (modified after Esser, 1982).

outstanding morphological feature of diatom cells is the wall (Fig. 20). The ecological significance of heavy silica walls and their peculiar appendages have been referred to above. The complete wall of one cell is called a **frustule**, and most (96%) of it is composed of quartzite (SiO_2), with an outer organic skin. The frustule is made up of two overlapping halves fitted together like a closed Petri plate. Although the frustule is extracellular in fully formed cells, the silica is actually laid down by the Golgi apparatus inside the cell membranes of developing cells. When the wall is fully formed, a new cell membrane is formed inside the wall, and the old membrane is shed. Apart from various spines and processes, the frustules of diatoms have very elaborate indentations which form beautiful symmetrical patterns. These are very pleasing to the eye, but their functions are unknown, except in the case of the **raphe** (Fig. 20). A raphe is a longitudinal fissure that occurs in the longitudinal axis of bilaterally symmetrical diatoms. Diatoms with a raphe appear to move (glide) by the secretion of mucilage from the cytoplasm through the raphe fissure.

The intracellular anatomy of diatoms is quite typically eukaryotic with chloroplasts, a single nucleus and a full complement of other organelles. The nuclei are probably diploid.

Planktonic diatoms are unicells, colonies or filaments (Fig. 6). The range of morphologies is greater than that found in any other group of phytoplankton.

In fast growing diatom populations reproduction occurs by division of each cell into two. The halves of the frustule are divided among the daughters so that each begins life with one ready-made frustule valve. The missing valve is produced inside the periphery of the valve inherited from the parent cell, so, of course, the cells get smaller as each generation begins.

Sexual reproduction is quite varied in diatoms. In all cases the gametes appear to form meiotically. There is variation in the types of gametes formed. In **oogamous** forms four uniflagellate male gametes are produced and released by each male cell. The sperms swim to the female cells that have undergone meiosis and produced an egg nucleus. The valve halves of the frustule of the female cell separate sufficiently to ensure that the sperm may gain entry. The zygote cell swells up to form, ultimately, a cell of normal maximum dimensions. Cell sizes are then reduced again through successive mitotic division.

It is interesting to note that the flagella of diatom sperms lack a central pair of microtubules (they are 9 + 0 in construction). The flagella are fully functional, so it is clear that, in this case, the central microtubules are not essential to coordinate swimming motion.

In other types of sexual reproduction in diatoms no sperms are produced. In some forms meiosis occurs and the zygote nucleus results from the fusion of haploid nuclei from the same parent cell. In another form of sexual reproduction gamete nuclei are transferred between parent cells that lie close to one another in pairs in a common mucilage.

Class Prymnesiophyceae. The tiny marine flagellates in this class are a major component of the phytoplankton in tropical waters, and probably of all oceanic communities.

The characteristic morphological feature of most cells of this group is the **haptonema** (Fig. 21). The haptonema is a coiled appendage, once

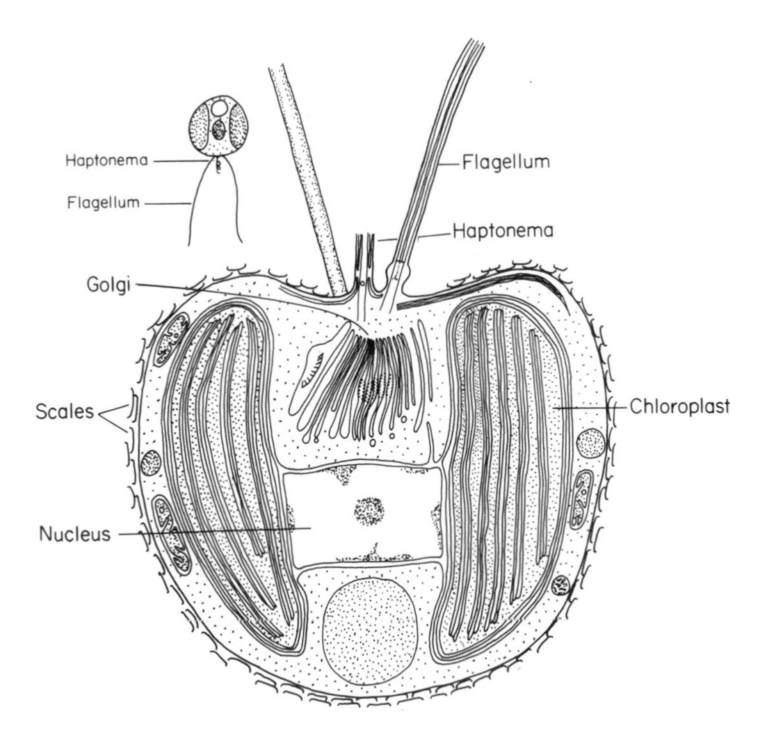

Figure 21. Fine structure of the cell of the Prymnesiophyceae (modified after Lee, 1980).

thought to represent a third flagellum. The internal anatomy of the haptonema is quite different from that of the flagellum (they differ in microtubule orientation, and in details of membrane organization).

The cell coverings of members of this class are quite unusual. Most forms are covered with scales (shown in section in Fig. 21). Some of the scales resemble armor plating (Fig. 6, *Coccolithus*) and are made up of calcium carbonate. In other forms the scales have long processes.

The flagella are usually paired and lack hairs or other appendages. Some forms have scales on their flagella. There are usually two chloroplasts and one nucleus in each cell. In types that have eyespots, the pigment granules are enclosed in the chloroplast membrane. Many of the cells in the Prymnesiophyceae are able to form pseudopodia. Phagocytosis is also common, thus members of the group have animal and plant characteristics.

Asexual reproduction occurs through zoospore formation. Sexual reproduction is isogamous and rare.

Class Chrysophyceae. The basic cell organization found in this class is shown in Fig. 22. The two flagella are of unequal length. The shorter is

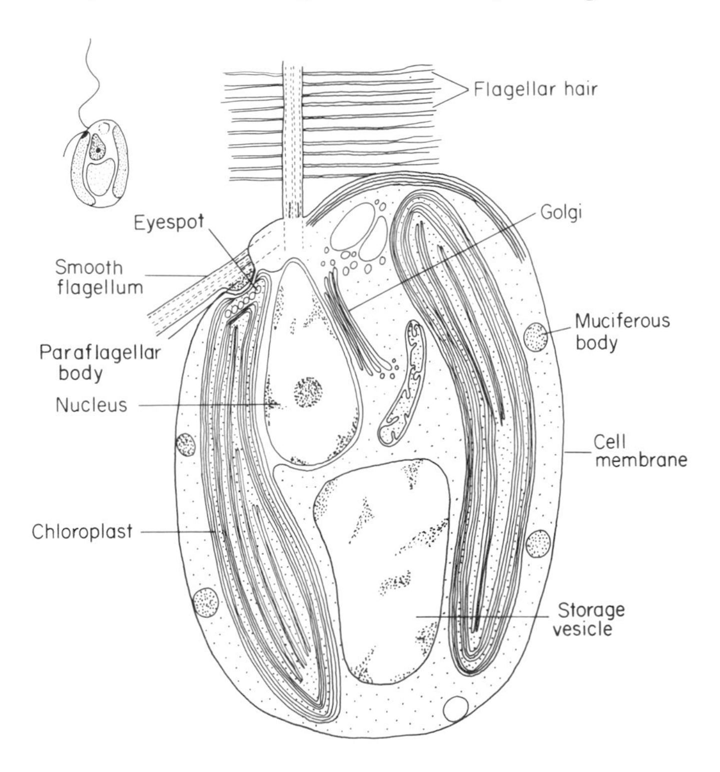

Figure 22. Fine structure of the cell of the Chrysophyceae (modified after Lee, 1980).

smooth and the longer bears hairs. The shorter flagellum bears a paraflagellar body at its base in alignment with the eyespot embedded in the cytoplasm.

Cell coverings are variable in this class. Some forms are bound only by the cell membranes. Scales are common and these are often silicified. Scales may bear spines, giving the cells a bristly appearance (Fig. 23). Purely organic scales also occur in the group. Some chrysophyceans secrete a **lorica** (Fig. 23), which is a flask-shaped structure within which the cells sit. One group of unicells, the silicoflagellates, has an elaborate internal skeleton of siliceous material.

Within chrysophycean cells there are usually only one or two chloroplasts. The thylakoids are stacked in groups of two or three. The cells are uninucleate.

Chrysophycean cells may be equipped with projectiles. The **muciferous bodies** shown in Fig. 22 discharge to form a fibrous network on the cell surface. The functions of the projectiles are unknown.

The diversity of morphologies is very great within the class (Fig. 24). Motile flagellate unicells have already been mentioned. In addition, there are non-motile unicells and colonies, motile colonies (flagellate) and ameboid forms. Filamentous and thalloid species also occur. Most planktonic members of the Chrysophyceae are simple unicells.

Asexual reproduction through flagellate spore formation is common. Sexual reproduction is rare and apparently isogamous.

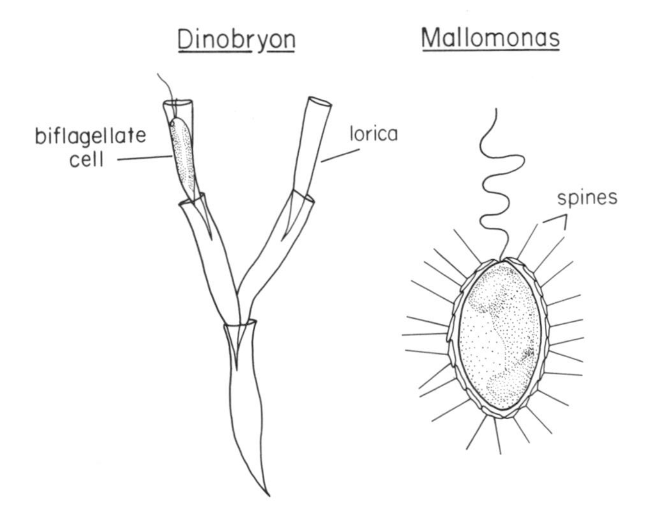

Figure 23. Two kinds of flagellate cells in the Chrysophyceae (from *Plant Diversity: An Evolutionary Approach.* © 1965, 1969 by Wadsworth Publishing Co.).

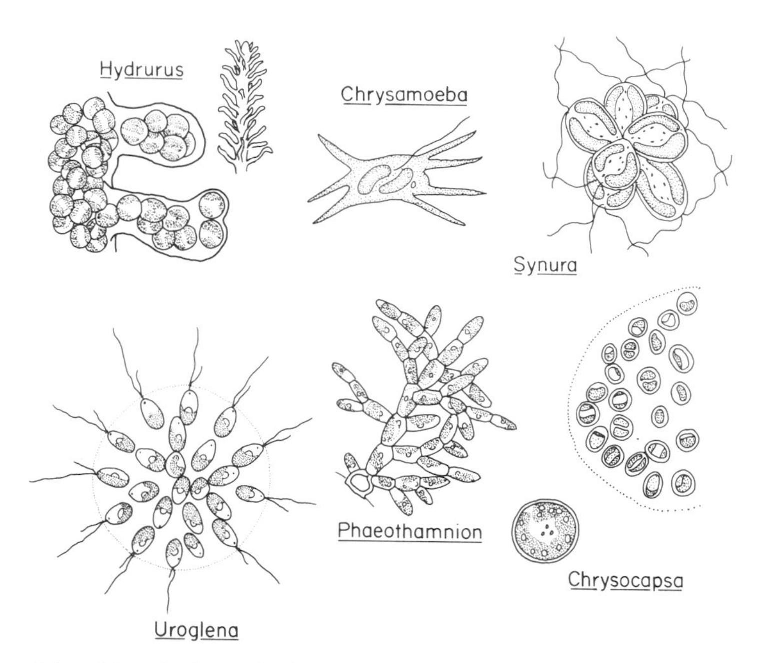

Figure 24. Morphological diversity in the Chrysophyceae (from *Plant Diversity: An Evolutionary Approach.* © 1965, 1969 by Wadsworth Publishing Co.)

Division Pyrrophyta

The members of this division are commonly called **dinoflagellates.** Several phytoplanktonic genera are illustrated in Fig. 6 (e.g., *Ceratium, Gymnodinium, Peridinium*). The cells of motile dinoflagellates are enclosed by the cell membrane beneath which there is a single layer of vesicles that normally contains cellulose plates (Fig. 25). Around the periphery of the cell there is a transverse groove called the **girdle**. A transverse flagellum fits within the girdle. This flagellum is twisted into a helical shape, and is covered with hairs. Another flagellum orginates in the girdle, but projects freely into the water.

The photosynthetic pigments are contained in small discoid chloro-plasts. The thylakoid membranes occur in stacks of three.

The nucleus of dinoflagellates is very unusual among the members of the Eukaryota. The chromosomes of most eukaryotic cells are composed of DNA and basic histone proteins. In the dinoflagellates the histone component is absent. Moreover, the chromosomes are permanently condensed and visible at all stages of the cell cycle. In most other eukaryotes the chromosomes can only be visualized during cell division. At other times they are dispersed.

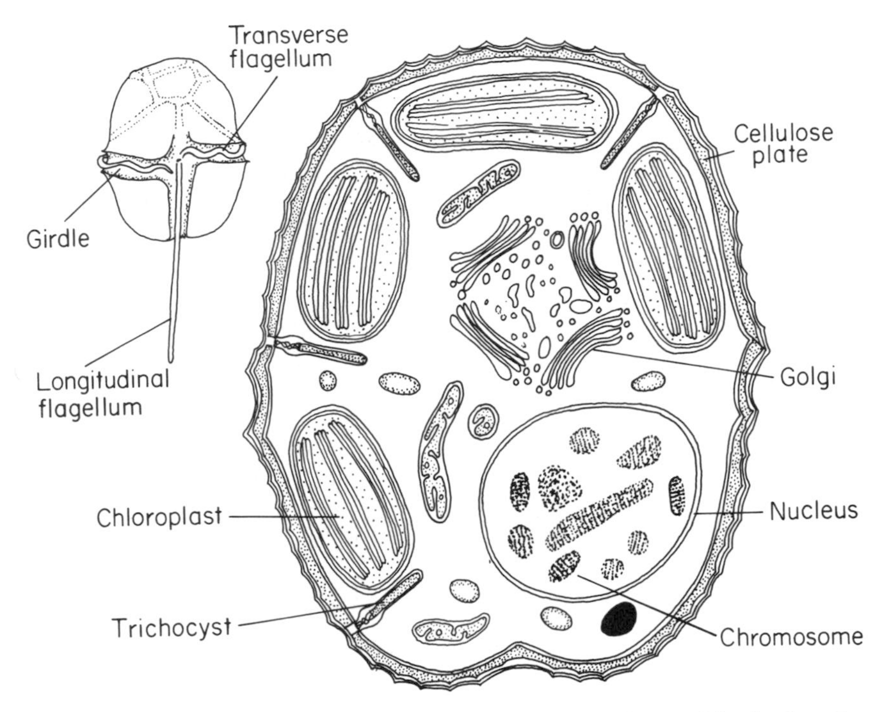

Figure 25. Structure of a motile dinoflagellate cell (modified after Lee, 1980).

Most dinoflagellates have projectiles near the cell surface that are fired if the cells are irritated. The commonest projectile is the **trichocyst** (Fig. 25) which discharges a 200μm long rod. **Cnidocysts** are explosive projectiles found in some groups of dinoflagellates. They are very similar to the nematocysts of sea anemones (Coelenterata), but their role in phytoplankters is unknown.

The morphological diversity of dinoflagellates (other than flagellate unicells) is shown in Fig. 26. Biflagellate unicells are the commonest morphological expression. There are also non-motile, and ameboid unicells. Motile colonial forms occur in the plankton, and filamentous benthic morphologies are known.

Most planktonic dinoflagellates are probably haploid and reproduce entirely by longitudinal cells division (asexual). Non-motile species produce motile flagellate spores. Sexual reproduction is quite rare and involves isogamous or anisogamous gametes.

Division Cryptophyta

The members of this group are commonly called **cryptomonads.** This is a very small group of flagellates. The cell membrane coated

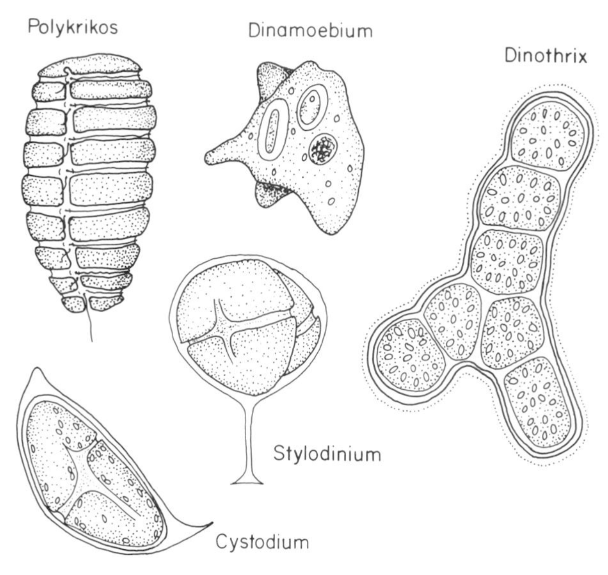

Figure 26. Morphological diversity in the dinoflagellates—colonies, non-motile unicells, ameboid cells and filaments (modified after Rabenhorst, 1933 and Fott, 1959).

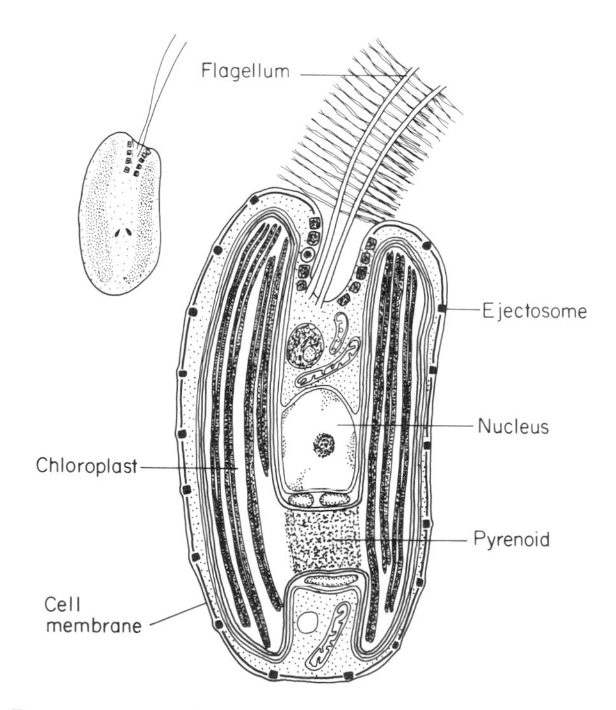

Figure 27. Fine structure of a cryptomonad cell (modified after Lee, 1980).

with fine fibers forms the outer cell layer (Fig. 27). Inside the cell there is a large two-lobed chloroplast and a pyrenoid. The thylakoids occur in stacks of two. The group is distinguished by the presence of water soluble **phycobilin** photosynthetic pigments (also found in the Rhodophyta and Cyanobacteria). The phycobilin occurs within the thylakoids and are not associated with phycobilisomes as they are in the Rhodophyta and Cyanobacteria.

The cryptomonads have intracellular projectiles called **ejectosomes.** These structures, of unknown function, are formed in the Golgi.

Reproduction in the group is by longitudinal cell division. Sexual reproduction is unknown.

4
Seaweeds

The predominant benthic vegetation component of rocky coasts is the seaweed flora. Seaweeds are macrophytic members of the following divisions: **Chlorophyta**, **Phaeophyta** and **Rhodophyta** (diagnosed in Table 4). There are microscopic benthic algae on the coastlines of the world, but these will not be considered here.

The lives of seaweeds are much more within the realm of human experience than those of phytoplankters. To begin with, seaweeds are macroscopic. In temperate waters many of the species are within the size range that human beings can perceive. When swimming or diving among submerged seaweeds we can feel and sometimes see the components of water motion that are important to macroscopic plants. This is in marked contrast to the phytoplankton environment. Phytoplankters are invisible in normal cell concentrations in the sea and they respond to water motion advection currents that are not felt by human sized swimmers.

In spite of living within the same size range as seaweeds, it is only very recently that we have begun to understand the functional significance of the types of seaweed morphologies that exist. To a large extent this understanding comes from the studies of Mark Littler and his associates (presently working at the Smithsonian Institution, Washington, D.C.)

Functional Forms of Seaweed Morphology

Mark and Dianne Littler have classified seaweed morphologies in six functional-form groups:

1. **Sheet group**—thin, tubular and sheet like.
2. **Filamentous group**—delicately branched (filamentous).
3. **Coarsely branched group**—more robust form of the filamentous group with a greater number of cell layers in thickness.
4. **Thick leathery group**—largest seaweeds with extensive tissue differentiation and rubbery or leathery texture.
5. **Jointed calcareous group**—seaweeds with heavy calcium carbonate impregnation, jointed and flexible at joints, upright.
6. **Crustose group**—flat crusts growing close to the substratum.

Representatives of all of these forms are shown in Fig. 28. It is important to realize that these diverse forms cut right across taxonomic and developmental boundaries. Thus for example, there are coarsely branched red, green and brown algae of quite separate evolutionary origins. The morphological convergences of separate evolutionary lines results from the common selective pressures that the coastal environment imposes on all seaweeds. The divergences among the functional form groups represent alternate strategies for dealing with three major sets of environmental variables:

1. **physiological and physical stress**,
2. **herbivory** and
3. **competition.**

It must be pointed out that adaptations to one environmental variable are often made at the expense of adaptations to another environmental variable. For example, the crustose morphology is a splendid adaptation to herbivory, but crusts are easily outcompeted by upright (foliose) forms of seaweeds. Thus there are costs and benefits associated with each morphological strategy. Some seaweed species have several morphologies in their life histories. This is called **bet hedging**.

The adaptations of each functional-form group to competition, herbivory and physical stress can be measured in terms of their growth rates, palatabilities and toughnesses, respectively. The results for a series of tropical seaweeds are shown in Fig. 29. The highest growth rates are to be found among the filamentous and sheet form species. Other groups are much less productive. The reason for this

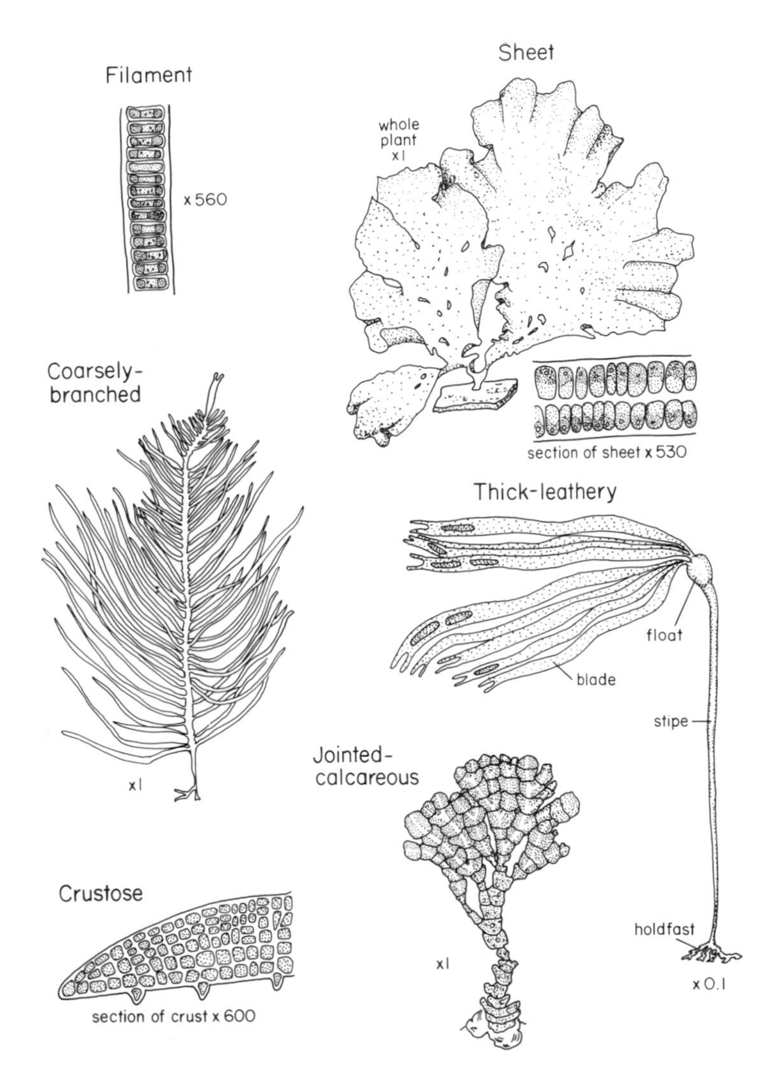

Figure 28. Six functional-form groups of seaweed morphology (from *Plant Diversity: An Evolutionary Approach.* © 1965, 1969 by Wadsworth Publishing Co.).

production differential is very simple. Filamentous and sheet forms are only one to a few cells in thallus thickness. This means that each cell is close to the raw materials for growth (light, mineral nutrients and dissolved inorganic carbon). These simple plants have high surface area to volume ratios and each cell is photosynthetic. This combination of characteristics leads to relatively high photosynthetic rates and a channeling of resources directly into new photosynthetic tissue. In contrast, other functional-form groups have relatively few

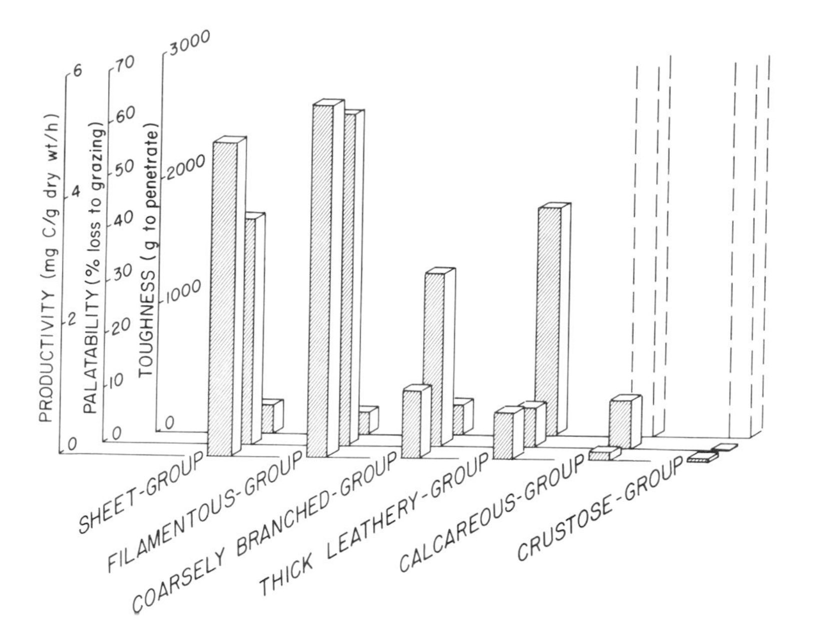

Figure 29. Photosynthetic production, palatability to herbivores and toughness of six functional form groups of seaweeds (modified after Littler et al., 1983).

photosynthetic cells in relation to non-photosynthetic cells. In the bulkiest forms only a small proportion of the tissues has access to the raw materials for growth. In addition, photosynthetic products in these latter forms are often channeled into storage compounds, rather than into structural compounds needed for growth.

The high productivity of filamentous and sheet forms ensure that they quickly occupy new space within communities. Moreover, it is now known that they will prevent colonization by other forms by preempting space. Sheets and filaments are excellent competitors. However, they are also very susceptible to grazing and environmental stress (as measured by palatability and toughness).

Crustose algae fall at the opposite end of the morphological spectrum from sheets and filaments. Crusts grow very slowly. They are very tough and, most importantly, very resistant to herbivory. In tropical locations where herbivore (mostly fish) grazing is very intense, algal crusts are the predominant marine benthic vegetation. If the grazing pressure is experimentally reduced, then the crusts are soon overgrown by other algal forms and they will die from light starvation. Thus crustose algae are poor competitors, but are well adapted to herbivory (indeed, herbivory is often required for their persistence). It is highly probable that many algal crusts have co-evolved with specific herbivores. In some cases a particular algal crust species chemically attracts and maintains its own specialist herbivore species.

Seaweeds and Water Movement

Seaweeds that are not calcareous have a density that is near that of sea water. Gravitational forces are therefore insignificant. Water flow is very important in the design of seaweed morphologies, and the property of flowing water that can cause failure of structural components is called **drag**. Seaweeds generally have a Reynolds number of more than 400, and above this level the most important component of drag is the **pressure gradient** around the organism. Drag is approximately proportional to the square of the velocity of the water flow. Drag is shown in graphical form in Fig. 30.

The velocity of water flow (and hence pressure drag) clearly varies with wind and tidal currents. Even at a constant surface current velocity of 1 m/s the flow rate will depend on position in the water column. As water moves over a rocky surface friction will slow its flow rate, but only noticeably within about 2 cm of the bottom. This narrow layer of slowly moving water is called the **boundary layer**. If water is moving at a surface velocity of 1 m/s, the flow rate will be reduced to 0.1 m/s at 2 cm above the bottom, and to perhaps 0.05 m/s at 40 μm above the bottom. All seaweeds spend some of

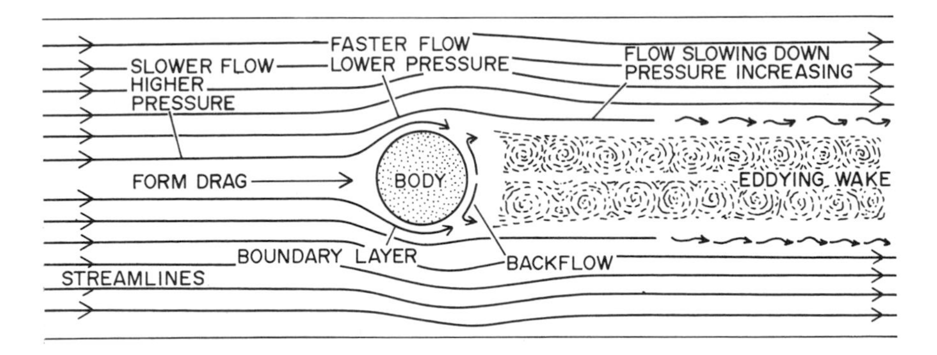

Figure 30. Flow forces around a seaweed thallus in a current . Behind the thallus an eddy wake separates from the mainstream flow so that the down-stream pressure is less than upstream pressure. This net downstream pressure is called **drag** (from Koehl, M.A.R. *The interaction of moving water and sessile organisms*. © (1982) by Scientific American, Inc.).

their lives (as spores, zygotes or germlings) in this slow moving water. Some species never extend beyond 2 cm in height, and spend all of their lives in the boundary layer. Crustose species are boundary layer occupants. Forms which extend up into the rapidly moving waters above must deal with enormous pressure drag forces and they do so by being **compliant** or **stretchy**. The alternate strategem to compliancy in a high pressure drag environment is to become very strong and rigid.

Either approach (elasticity or rigidity) increases the energy required before breakage occurs. The breaking stresses of seaweeds are very low, but breaking extension is very high so that the mean work needed to break flexible seaweeds is similar to cast iron or wood. The flexibility of seaweeds enables them to employ various pressure drag avoidance movements. First, it is known that, as water velocity increases, plants bend over closer to the sea bottom and into the slower moving water layers. Second, since most seaweeds live in an oscillating surge zone with water moving back and forth, they can move with the water flow. The water will not move much with respect to the plants until they are fully bent over.

The largest seaweeds (>30 m long) do not live in an oscillating surge zone. Oscillating flow occurs in shallower waters than those in which the giant kelps live. These enormous brown algae have floats which take their distal portion up to the surface where current flow is unidirectional. The surface canopies act as light harvesting devices. Below the canopy illumination is reduced to a small fraction of surface levels, and plant growth is stunted. The simplest of the giant seaweeds (*Nereocystis* in Fig. 28) consists of (a) the light harvesting blades, (b) the float which holds the blades at the surface, and (c) the stipe (stem) which anchors the distal portions to (d) the holdfast which anchors the whole structure to the substratum. Each of these structures may be regarded as a plant **organ**. Organ differentiation is common among seaweeds. Even the simplest filaments are differentiated into (a) a holdfast cell, and (b) all of the remaining photosynthetic cells of the thallus.

Developmental Diversity of Seaweed Thalli

Convergent evolution has led to the development of common functional forms of quite different developmental origins. For example, there are red and brown seaweeds in the **thick-leathery** functional-form group. The red and brown algae in this group have broadly similar functional attributes with respect to stress, competition and herbivory. However, the developmental origins of bulky red and brown seaweeds can be totally different.

Seaweeds begin their lives as unicells (spores or zygotes) and development proceeds with the formation of a tube (usually called a germ tube) on one side of the roughly spherical spore or zygote. As the tube elongates cell division may occur at right angles to the axis of elongation. This produces a filament. The simplest seaweeds remain filamentous through their lives (Fig. 28). In some green seaweeds there may not be cell division at right angles to the axis of germ tube elongation. Organelle division occurs, but a large structure without cross walls can develop (Fig. 31). This is called **coenocytic** construction. In advanced seaweeds a bulky plant body is achieved through the development of either **parenchyma** or **pseudoparenchyma**. Parenchyma is produced by repeated cell division in a variety of planes. A very simple parenchyma is found in the brown alga *Sphacelaria* (Fig. 32). In *Sphacelaria* the cells in the various parenchyma layers that are produced are not differentiated, that is, they all perform quite similar functions. In the complex parenchymatous forms like *Nereocystis* (Fig. 28) the cells that are produced in various layers undergo extensive differentiation. Only the outermost cell layers are pigmented and photosynthetic. Within the pigmented layer is a tissue of colorless box shaped cells called the **cortex** (Fig. 33). In the middle of the thallus is a central **medulla** of very elongated cells (Fig. 33). The young **sieve filaments** in the medulla are known to have a translocatory role. Within the cells of this tissue soluble polysaccharides and amino acids are transported from the photosynthetic organs (blades) to the growing parts of the thallus.

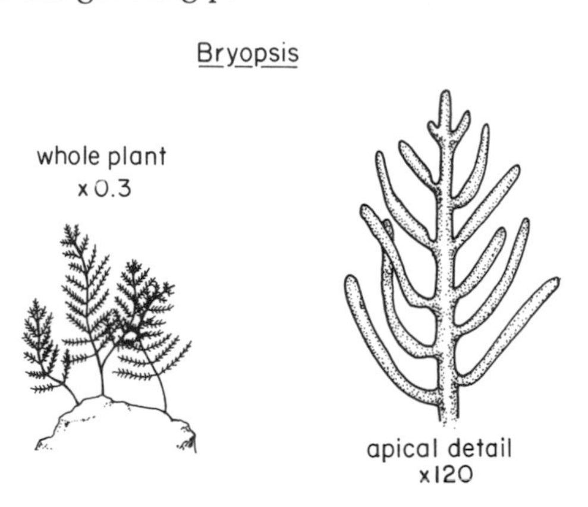

Figure 31. Coenocytic thallus construction in *Bryopsis*, a green alga (from *Plant Diversity: An Evolutionary Approach.* © 1965, 1969 by Wadsworth Publishing Co.).

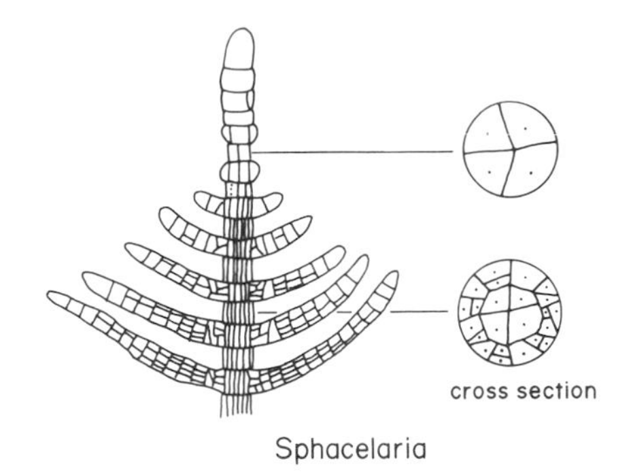

Figure 32. Simple parenchyma formation in *Sphacelaria*, a brown alga (modified after Chapman, 1979).

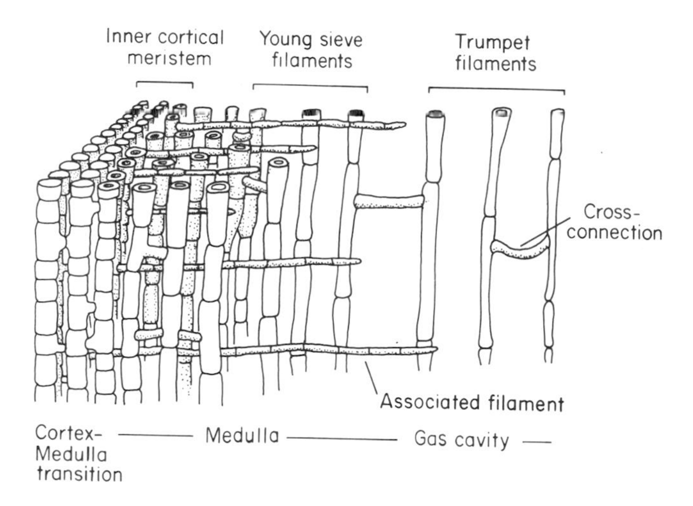

Figure 33. Medullary and cortical cell differentiation in the giant kelp *Nereocystis* (modified after Nicholson, 1976).

Most of the red algae of bulky construction are made up of pseudoparenchyma. Pseudoparenchyma is derived from the coalescence of filaments. In very simple pseudoparenchyma this is easy to see (Fig. 34). Notice that in *Hypoglossum* (Fig. 34) cell division in the webbed thallus occurs in a restricted number of planes (compare this with *Sphacelaria*, Fig. 32). In the bulkiest of red algae the filaments at various depths in the thallus differentiate and it is often impossible to identify pseudoparenchyma in the mature plant body. Nevertheless, the developmental basis of the construction is filament coalescence.

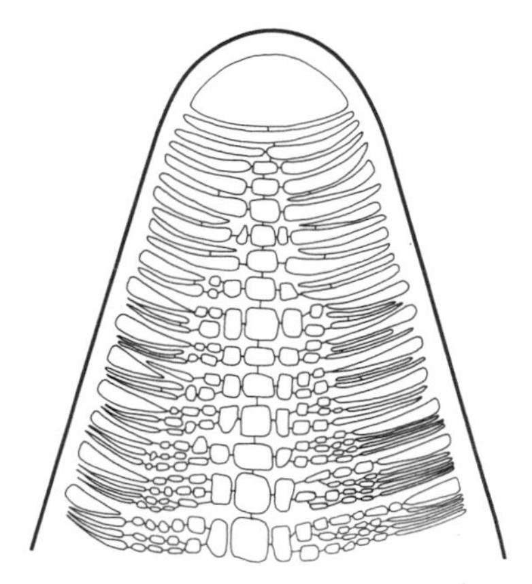

Figure 34. Simple pseudoparenchyma in the red alga *Hypoglossum* (modified after Chapman, 1979).

Reproduction

Seaweeds reproduce by **vegetative**, **asexual** or **sexual** processes. The vegetative process most commonly employed is **fragmentation**. Fragmentation occurs when part of the parent plant breaks away and reforms a new individual. In some cases the vegetative fragments are specialized differentiated propagules, but these are relatively uncommon. Asexual reproduction involves the formation of **spores** within a parent cell wall. The spores are released by dissolution of the parent cell wall. In no case is any part of the parent cell wall incorporated into the spore, and this attribute separates asexual reproduction from vegetative propagation. Nearly all spores in green and brown seaweeds are flagellated **zoospores** which disperse in the plankton under their own swimming power. In red algae spores are always nonmotile, and must disperse passively with water currents.

Sexual reproduction is widespread among seaweed groups and is the only form of reproduction in some brown algae. The product of sexual reproduction is the zygote. The zygote results from the fusion of two haploid gametes. Seaweed gamete fusion (**syngamy**) may be **isogamous**, **anisogamous** or **oogamous** (female gamete is a large, nonmotile egg). In most green and brown seaweeds the gametes (male and female) are released into the water like spores. Fertilization is external. In the red algae only the male gametes are released. The egg nuclei remain in the **oogonia** (egg-producing cells) on the female parent plant. Red algal male gametes must be carried by water currents to the oogonia. Fertilization occurs when the male nucleus is injected into the oogonium.

In the simplest seaweeds **sporangia** (spore-forming cells) and **gametangia** (gamete-forming cells) are not differentiated from other cells in the thallus. In complex forms reproductive cells are produced in special multicellular structures (described by a complex terminology that is not used here).

In green and brown seaweeds the zygote develops directly into another organism through cell division (the first divisions may be mitotic or meiotic). In most red algae there is a complex sequence of post-fertilization events. Basically the zygote nucleus divides many times to produce numerous cells from the original nucleus. Most of these cells develop into separate spores. The spores are released and each has the capacity to produce a new plant. Thus each fertilization event may lead to the production of thousands of genetically identical offspring. In green and brown seaweeds each zygote nucleus normally produces one individual. The differences in post-fertilization behaviors among the seaweed groups may relate to the motility, or lack of motility, among male gametes. It is now known that in many brown algae female gametes release a water soluble hormone that attracts male gametes. The males swim toward and easily locate the females. In red algae the male gametes cannot move actively to the females and therefore the chances of fertilization must be rarer than in other eukaryotic algal groups. It seems reasonable to propose that multiple spore production from single zygotes of red algae is an adaptation to the rareness of fertilization.

In the simplest seaweed life histories spores and zygotes develop into plants that are similar to the parent that produced them. In **heteromorphic** life histories the spores and zygotes grow into plants that do not resemble their parents. Extreme heteromorphism occurs among the giant kelps (brown algae) like *Nereocystis* (Fig. 28). These giant plants produce spores that grow into minute sexual plants. It seems most likely that in the wild the female sexual plant consists of only one cell and the male of only a few cells. These microscopic structures produce eggs and sperms, respectively. The sperms are released when a chemical message is received from a nearby mature oogonium. The sperms swim down a hormone concentration gradient to the female and effect fertilization. The zygote grows into a new kelp.

Less extreme heteromorphism is common among the seaweed groups. Very often heteromorphic phases occur in different seasons of the year. In the mid and high intertidal zones of temperate North America several species of macroscopic ephemeral plants commonly occur only in winter. The species pass the summer as microscopic plants. Recent work has shown that this seasonal alternation may represent an adaptation to seasonal variation in herbivore pressure. The major snail herbivores are very active in summer, and much less so in winter. If the herbivores are experimentally removed in summer, many species' macroscopic phases will survive the period when they are normally absent. Under usual circumstances the microscopic phases represent mechanisms for surviving elevated grazing pressure during the summer.

In many groups of seaweeds the microscopic and macroscopic life history phases may have a similar seasonal occurrence. In this instance the microscopic phase is seen as a mechanism for surviving the patchy variation in grazing pressure. The macroscopic phase, on the other hand, is regarded as an adaptation in the competition for light that occurs between plants. In the absence of grazers the microscopic plants are soon overgrown and eliminated by larger seaweeds. A macroscopic phase can survive this competitive form of biotic interaction.

In nearly all seaweeds with heteromorphic or isomorphic life histories the two free-living phases have different ploidies. The commonest combinations of ploidies and morphological phases are shown in Fig. 35. There are many variations on the themes shown. For example, male and female plants (both haploid) may be of quite different sizes and morphologies.

The significance of differences in ploidies among the life history phases is not at all clear. In the red alga *Chondrus crispus* (Irish Moss) haploid male and female plants are morphologically similar to the diploids. The cell wall chemistries of diploid and haploid plants are slightly different, so it is possible to identify the ploidies of non-reproductive plants. Work has shown that in the intertidal zone of the Canadian Maritime provinces haploids make up by far the greatest proportion of the populations (>85%). In another alga, *Plocamium caritilagineum*, the vast majority of reproductive plants in European populations are diploid. The adaptive significance of these differences in ploidy ratios are unknown.

Systematic Diversity of Seaweeds

Seaweeds belong to the divisions Phaeophyta, Chlorophyta and Rhodophyta of the kingdom Eukaryota (see Table 4 for a diagnosis). Partial systematic surveys of the Chlorophyta and Rhodophyta are given in the previous chapter.

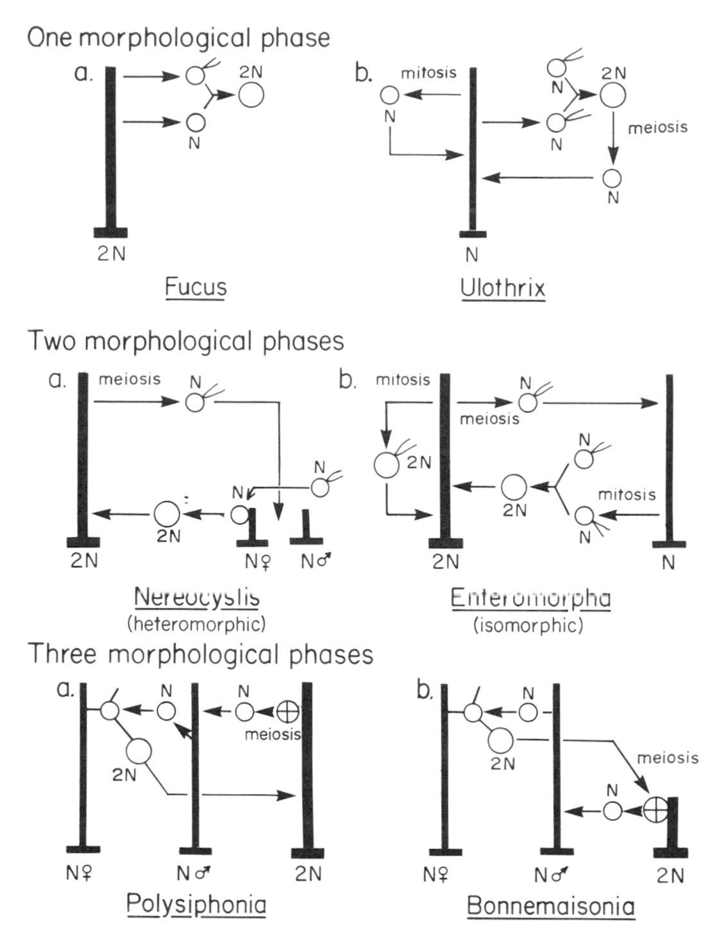

Figure 35. Combinations of morphological phases and ploidy levels found in seaweed life histories.

DIVISION PHAEOPHYTA

Virtually all the brown algae that belong to this division are seaweeds. There are no phytoplankters, apart from reproductive bodies (spores and gametes). Very few species occur in fresh water.

Vegatative cells of brown algae are sheathed in a cell wall. The wall has two carbohydrate components. One phase is fibrillar cellulose (glucose polymer). The other is amorphous and mucilaginous and made up of **alginic acid** and **fucoidin**. Alginic acid is a copolymer of two uronic acids (mannuronic acid and guluronic acid). Fucoidin is a polymer of the sugar fucose. The amorphous carbohydrates are commercially exploited. Alginic acid is used as a gelling agent in various foods and in the cosmetic industry. In live seaweeds alginic acid polymer structure is variable. This variability determines the flexibility of plants.

Inside the cell walls of brown algae the fine structure is similar to that of the Chrysophyta. The photosynthetic pigments are contained in the chloroplast thylakoids which are stacked in threes (Fig. 36).

Physodes are especially prominent in brown algal cells. These structures contain phenolic substances in solution. The phenolics apparently inhibit the growth of microalgae. They may also inhibit grazing. Thus physodes may be regarded as chemical warfare devices. Physodes are produced from chloroplasts.

Zoospores and gametes of brown algae bear flagella. The flagellar structure is similar to that of the Chrysophyceae (Fig. 22). The morphological organization of the brown algae is diverse (Fig. 37), ranging from microscopic filaments to giant kelps. There seem to be several partially distinct developmental lines (Fig. 37). The simplest growth type is an undifferentiated filament in which all cells perform similar functions. Differentiation of filaments occurs through the localization of cell division (in the meristems) and through branching. Pseudoparenchyma derived from filament coalescence is seen in *Leathesia* (Fig. 37) which has apical cell division. Very simple parenchyma is found in *Sphacelaria, Dictyota* and *Syringoderma* all of which have apical meristems. In the massive kelps like *Laminaria* and *Macrocystis*, cell division produces many differentiated cell layers or tissues. In the laminarians cell division is restricted to the base of the blade.

There are two types of reproductive structures involved in the production of spores and gametes in the Phaeophyta These are the

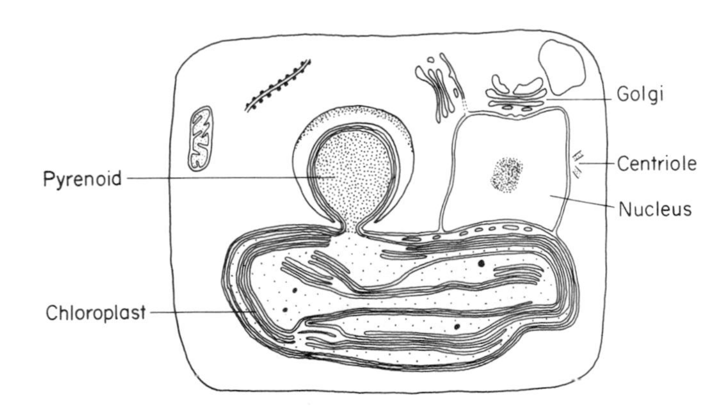

Figure 36. Basic fine structure of a brown algal cell (wall not shown) (modified after Bouck, 1965).

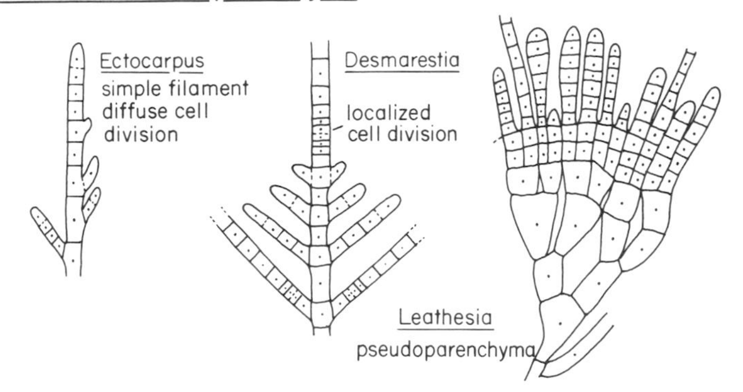

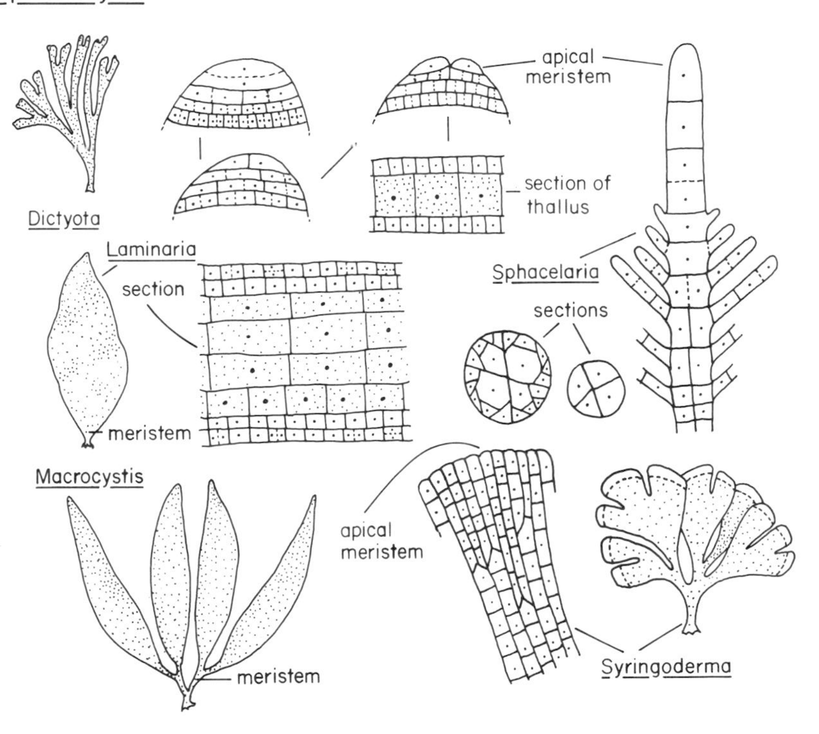

Figure 37. Range of developmental morphologies in brown algae. Filaments and derived filamentous forms (pseudoparenchyma), true parenchyma (from *Plant Diversity: An Evolutionary Approach.* © 1965, 1969 by Wadsworth Publishing Co.).

unilocular and **plurilocular** sporangia (Fig. 38). The plurilocular sporangia are many chambered (celled) and each chamber produces reproductive structures by mitosis. Unilocular sporangia consist of single cells that undergo meiosis to produce haploid swarmers. In the

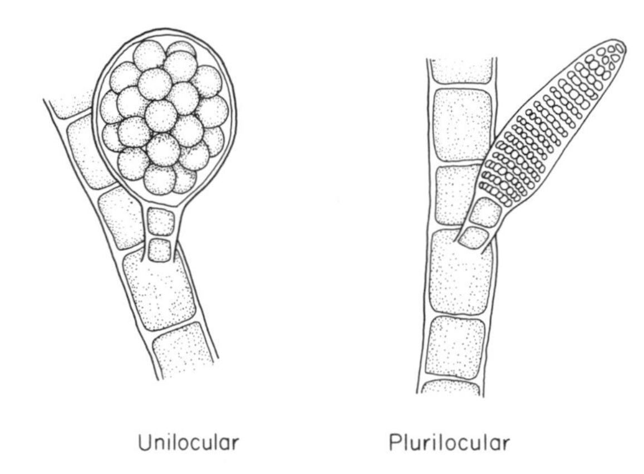

Figure 38. Meiotic (unilocular) and mitotic (plurilocular) sporangia of the brown alga *Ectocarpus* (from *Plant Diversity: An Evolutionary Approach.* © 1965, 1969 by Wadsworth Publishing Co.).

vast majority of brown algae the swarmers produced mitotically (in plurilocular sporangia) from haploid plants act as gametes. However, the diploid spore products of plurilocular sporangia on diploid plants act only as spores. In most cases the haploid products of meiosis in unilocular sporangia act only as spores to produce a haploid morphological phase. The unilocular meiotic sporangia of *Nereocystis* are found on the macroscopic plant as shown in Fig. 35. Modified plurilocular sporangia (acting as gametangia, because they produce gametes) are found on the microscopic stages.

The life history of *Fucus* (Fig. 35a) is unusual among plants because there is only one (diploid) morphological phase. The diploid plants produce haploid gametes from modified unilocular sporangia. In the other types of life histories in brown algae the spores from unilocular sporangia do not act as gametes. Some authors have claimed that *Fucus* and its relatives do have two morphological phases in their life histories. The gametes are seen as separate haploid phases. This interpretation has not received wide acceptance.

The life histories of *Fucus* and *Nereocystis* are shown in Fig. 35. Many other brown algae have isomorphic life histories similar to that of *Enteromorpha* (a green alga) which is also shown in Fig. 35.

DIVISION RHODOPHYTA

The cellular organization of this division was reviewed briefly in the last chapter and will not be repeated again here. Within the division as a whole there is a great diversity of morphological organizations. The simplest type is unicellular and non-motile. In addition, there are filamentous forms, both branched and unbranched, with apical localization of growth (meristems) or with diffuse cell division. Parenchymatous forms are very simple (Fig. 39). The greatest complexity of form is seen in pseudoparenchymatous constructions with considerable tissue differentiation (e.g., Fig. 40). Nearly all the pseudoparenchymatous types have apical growth in each of the filaments that coalesce to form the thallus.

Sexual reproduction is common among the more complex red algae. It is quite rare or absent among filamentous forms. The basic features of gamete formation have been described above (pp. 51-2).

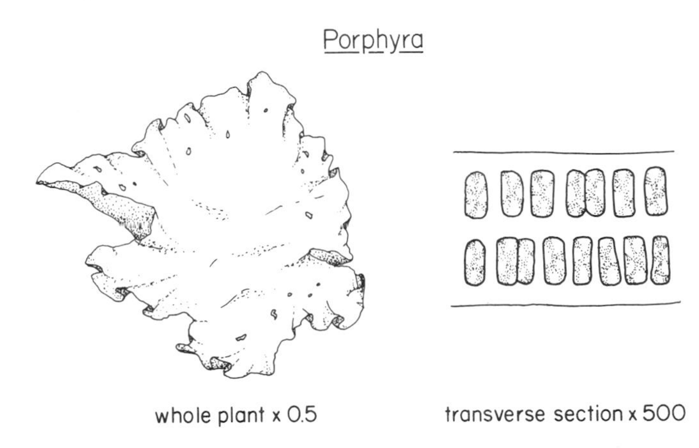

Figure 39. Parenchymatous construction of *Porphyra*, a red alga (from *Plant Diversity: An Evolutionary Approach.* © 1965, 1969 by Wadsworth Publishing Co.).

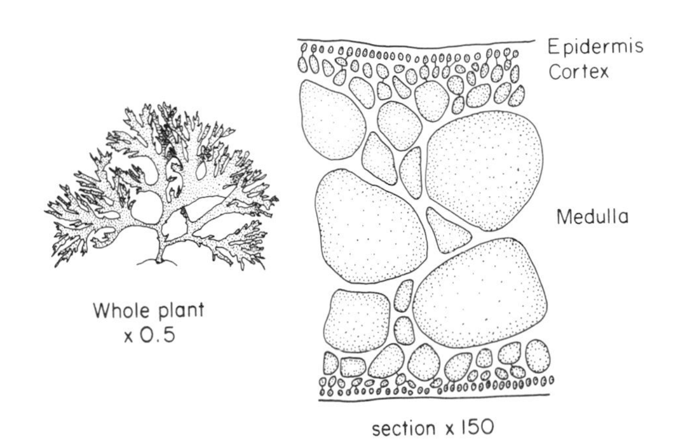

Figure 40. Tissue differentiation in the pseudoparenchymatous red alga *Callophyllis* (from *Plant Diversity: An Evolutionary Approach.* © 1965, 1969 by Wadsworth Publishing Co.).

Sexual fusion leads to the formation of a zygote nucleus that divides mitotically to produce a multicellular mass (which is diploid) that is attached to the haploid plant. The diploid cell mass is often regarded as a separate morphological phase. The diploid spores that it releases grow into free-living morphological phases that produce spores meiotically (Fig. 35, *Polysiphonia* and *Bonnemaisonia*). The haploid spores produced by the free-living diploid phase develop into gametangial phases. The great majority of asexual spores of red algae are produced in this way.

DIVISION CHLOROPHYTA

The cellular organization of this division was reviewed in the last chapter. During the 1970s the cell structure of this group came under close scrutiny and a new viewpoint has evolved. According to most recent work the green algae fall into two groups depending on their cellular organization. Radical proponents of the new view suggest that the two groups of green algal cell structures should lead to a massive reorganization of the taxonomy of the group. This is of no concern here, but it is necessary to provide a brief description of the two types of cell structure. First of all, in the so-called **chlorophycean line** the flagella of motile cells are attached at the anterior end of the cell with four cruciately arranged flagellar roots (proteinaceous tubular structures, Fig. 41). In the **bryophytan line** the flagella are inserted laterally and are associated with a single microtubular band (Fig. 41).

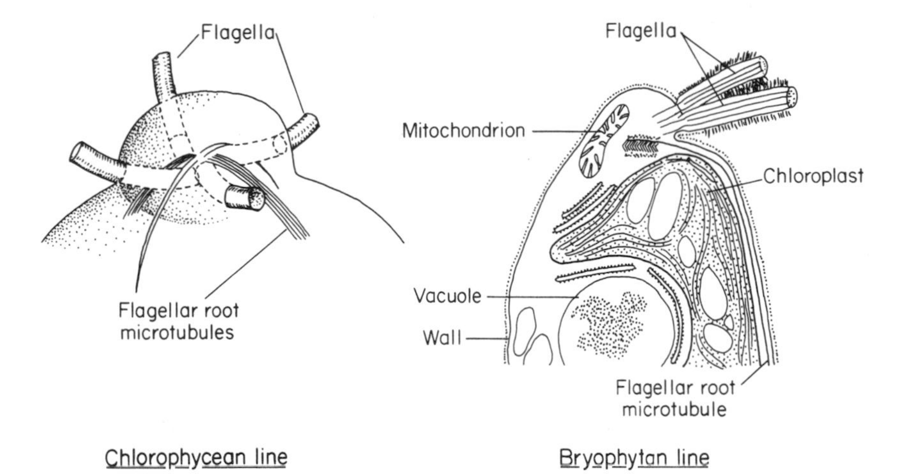

Figure 41. Arrangement of flagella and flagellar roots (microtubules) at the anterior ends of "chlorophycean-line" and "bryophytan-line" green algae (modified after Hoek, 1981).

Nuclear division is different in the two lines. In the bryophytan line there is a classical pattern with the development of a long-lived nuclear spindle (made up of microtubules) and the formation of a new cross wall from the coalescence of the Golgi vesicles between spindle microtubes (Fig. 42). In the chlorophycean line the mitotic spindle disperses very shortly after nuclear division, so the two daughter nuclei remain close together. A new cross wall develops along a new set of microtubules formed between the daughter nuclei at right angles to the spindle microtubules (Fig. 42). Among the green algae that are recognized as seaweeds only certain coenocytic forms have characteristics of the bryophytan line of cell organization. The features that distinguish the bryophytan line are also shared by vascular plants and bryophytes.

Seaweeds in the Chlorophyta have less morphological diversity than the Phaeophyta or Rhodophyta. Most forms of benthic green algae in cold-temperate waters are filaments or pseudoparenchyma. In the tropics the coenocytic forms are a conspicuous component of the flora. Apart from the simpler forms such as *Bryopsis* (Fig. 31), coenocytic algae have diversified into complex pseudoparenchyma (Fig. 43) and giant unicells like *Acetabularia* (Fig. 43). Some of the pseudoparenchymatous coenocytes (e.g., *Halimeda*) are calcified and jointed.

Among the morphological diversity of green seaweeds there is a very rare Australian **palmelloid** form called *Palmoclathrus* (Fig. 44). A palmella consists of non-motile cells embedded in a common mucilage. The cell number and disposition is indeterminate. In *Palmoclathrus* the undifferentiated cells are embedded in a stiff common mucilage.

Among the green seaweeds there may be one or two morphological phases in the life histories. Where there is only one, plants may be diploid or haploid (Fig. 35). *Codium* has a life history similar to that of *Fucus* (Fig. 35) in which gametes are produced by meiosis from

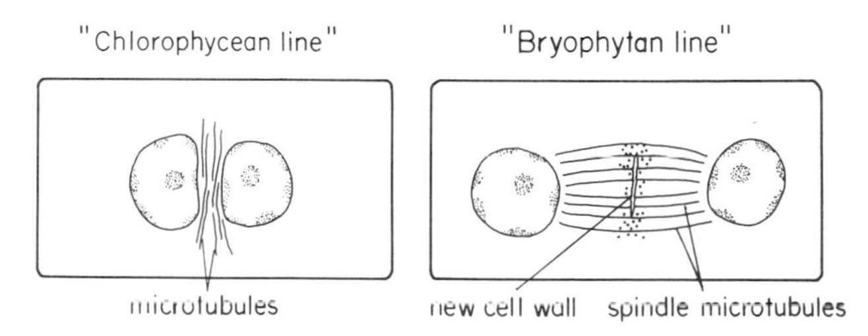

Figure 42. Types of cell division in "chlorophycean-line" and "bryophytan-line" green algae (see text for details) (modified after Lee, 1980).

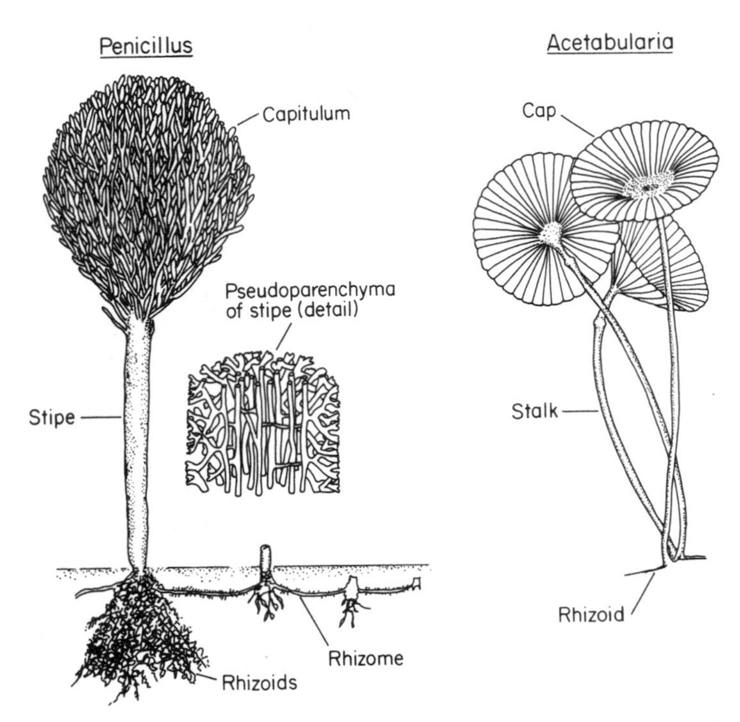

Figure 43. Elaborate coenocytic forms of green algae (modified after van den Hoek, 1981).

Figure 44. *Palmoclathrus*—a palmelloid green seaweed (drawn from a photograph by Womersley, 1970).

diploid parents. *Ulothrix* (Fig. 35) plants are haploid and meiosis occurs in the zygotes.

Where there are two morphological phases in the life history, diploids and haploids may be isomorphic (*Enteromorpha*, Fig. 35). Alternatively, the diploid phase (e.g., *Acrosiphonia*) or the haploid phase (e.g., *Derbesia*) may be very reduced, making the life history heteromorphic.

Asexual reproduction by motile spores is found in many green seaweeds. Sexual reproduction may be isogamous, anisogamous or oogamous and always involves motile (flagellate) male gametes. Fertilization is nearly always external.

5

Terrestrial Conditions and Plant Life

It seems most probable that the vascular plants and bryophytes that dominate the modern terrestrial flora had aquatic ancestors. These ancestors were likely green algae. For an algal ancestor the terrestrial environment must have been a daunting prospect for colonization. The problems that had to be overcome were: (a) mechanical support and (b) drought and problems associated with water supply.

Mechanical Support

Attached algae living in turbulent waters are subjected to mechanical stress. Most of this stress is in the form of tension as the current pulls on the plants. However, the density of algae is similar to that of water so that support systems are not essential. When large seaweeds are removed from water they collapse because their density is much greater than that of air. The portions of terrestrial plants that are in the atmosphere require a support system. Plants portions that are underground have no such requirement. However, roots are subjected to tensile and compression strains caused by wind pressure on the aerial plant portions.

The force of gravity is chiefly a threat to large plants. It is important to remember that plant volume increases as the cube of the linear dimension. Hence the mass will increase disproportionately with increasing plant height. It is therefore not surprising to find the development of massive support systems in the largest of living organisms, terrestrial trees. Failure of aerial support system materials leads to compression and buckling (compression is the reverse of tension). Pure tension or compression forces are rather uncommon states of stress. By far the commonest situation is bending. For terrestrial plants bending can occur under the force of gravity (in a horizontal tree limb) or wind. These concepts are illustrated in Fig. 45.

In gale force winds the drag forces on a tree may be greater than its own weight. Bending exerts greater stresses in plant stems than simple compression of the same force. Uprooting occurs when the total bending force exerted by the trunk on the root system exceeds the maximum restoring movement that the surrounding soil can exert on the roots. For rigid objects the drag force that causes bending increases with the square of the wind velocity. Uprooting is only one consequence of extreme bending. Stem breakage can also occur, and is a serious problem for cereal growers, especially when strong winds

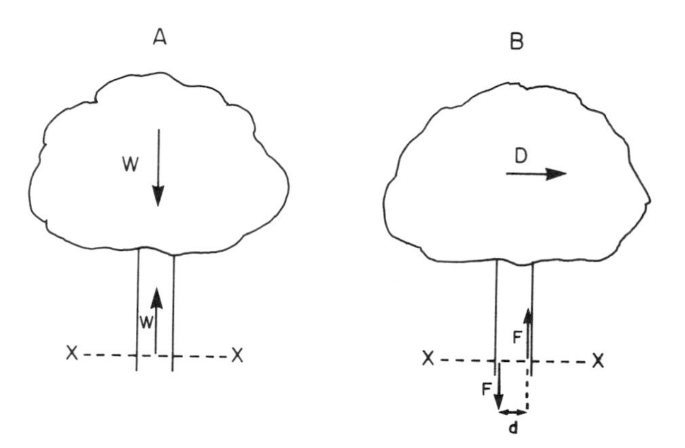

Figure 45. Forces acting on above ground parts of a tree. A. In still air the wood of the trunk is compressed by the weight of the tree so that the part of the trunk above XX exerts an upward force equal to the weight above. B. In moving air the trunk is stretched on the windward side and compressed on the lee side so that the parts below XX exert on the parts above XX the forces F which balance the movement due to the form drag D (modified after McNeill Alexander, 1971).

are combined with heavy rainfall. In natural winds the magnitude of the forces oscillates with the successive passage of eddies in the surface wind. Individual rice plants have been shown to respond to this oscillation by swaying. A constant period of sway occurs once a critical starting velocity has been achieved.

Drought and Problems Associated with Water Supply

Some of the algae dealt with in previous chapters are subjected to drought. For example, many high intertidal seaweeds are exposed to desiccating conditions for several days at a time. Some seaweeds do tolerate drought, but they do not avoid it. The difference between drought **tolerance** and drought **avoidance** is not trivial. The tolerators can lose more than 90% of tissue water during desiccation. Most metabolic functions are almost shut down under these conditions. There is no net photosynthesis, and hence no growth. The metabolism of drought tolerators recovers quickly on re-immersion. Drought avoiders maintain their metabolic activities during periods when external water is unavailable by restricting water loss. All vascular plants are drought avoiders. When water is available to drought avoiding plants on land it must be obtained from the soil. Water in the atmosphere is unavailable because of drought resistant anatomies of vascular land plants. The problem of drought on land is therefore mainly a question of soil water availability. Drought avoidance anatomies also influence mineral nutrient availability. Only minerals in the soil can be used for the growth of land plants.

How much of the earth's water is available to the roots of plants? More than 97% of the water on the planet is in the oceans. Of the remaining 3%, two-thirds is locked up in snow caps and glaciers. Only 0.6% of the earth's water occurs on land and most of this is in ground water that is too deep for plant roots to reach. Only 1% of the water on land occurs in soil that is shallow enough to allow plant access. This means that only 0.006% of the earth's water is available for plants on land. It is therefore hardly surprising that only 10% of the surface of the land is suitable for the growth of crops.

The availability of water to plants is best examined by using the concept of **water potential**. Water potential is a way of expressing the status of water in thermodynamic terms. According to the laws of thermodynamics systems will move spontaneously in directions that decrease potential energy. Water potential (Ψ) is a measure of the free energy of water (in plants, in soil solutions, etc.) in comparison to the free energy of pure water. The water potential of a system indicates the tendency of pure water to move toward or away from that system. If we are comparing two systems, neither of them pure water, their relative water potentials indicate the direction in which water will move spontaneously.

Pure water has been arbitrarily assigned a water potential of zero bars. A **bar** is a pressure unit (1 bar = 0.987 atmospheres = 10^6 dynes/cm^2). The chemical energy of water in the soil and in plants is lower than pure water, so that water potential is expressed in negative numbers. A soil's ability to hold water is described in terms of water potential in the following equation:

$$\Psi_{soil} = \Psi_m + \Psi_\pi + \Psi_p + \Psi_g$$

Ψ_m is the matrix potential which can be thought of as the force with which water is held by capillary action and adsorption to the soil colloids. This force always reduces the free energy of water and is therefore negative. Ψ_π is that part of the soil water potential represented by the osmotic pressure of the soil solution (negligibly small). Ψ_p is the component of soil water potential represented by hydrostatic pressure of the water in the pore space (negligibly small). Ψ_g is the gravitational potential of soil water and is only significant in saturated soils. Thus under normal circumstances soil water potential is primarily determined by capillary and adsorption forces. The forces holding water in the soil are stronger (Ψ_{soil} becomes more negative) as the soil dries.

A plant can only withdraw water from soil when the water potential of its roots is more negative that the water potential of the soil solution ($\Psi_{soil} > \Psi_{root}$). In straighforward terms this means that the osmotic potential of cell sap in roots has to give an adequate water potential to overcome the forces (capillary and adsorption) holding water to soil. Under drought conditions Ψ_{soil} increases so that the plants are unable to take up water. It should also be remembered that below $-1°C$ all capillary water in soil is frozen and cannot enter plants. Soil temperatures below $-1°C$ are normal winter conditions for huge tracts of the earth's surface, so that drought is not simply restricted to the hotter regions of the planet.

Drought is associated with the extreme temperature range that occurs on land. In the waters of the oceans temperatures range from about $-1°C$ to about $30°C$. These are conditions that pose no danger to the vital functions of plants. On land there are regions where temperatures never rise significantly above freezing point and, as plants have no temperature regulation, primary production is prohibited. At low latitudes air temperatures approaching 60°C have been measured. Normally high temperatures in the tropics are not a threat to plant life. However, evaporation of soil moisture increases with temperature and this leads to drought (unless replenishment occurs).

Drought conditions promote lightning fires. Lightning is the major cause of natural fires. Fire is clearly an aerial phenomenon never experienced by marine plants. Plants in most terrestrial communities have had to evolve with fire as a selective force of some importance. It must be remembered that during a fire the temperature of the soil will not rise above 100°C until all the water has evaporated. This evaporation requires considerable energy. Generally a light surface fire in a grassland heats only the top fraction of a centimeter of soil to 100°C. Hence plants may survive as underground portions. Some dry grassland species even seem to require fire as an agent to promote seed germination and fire is often an important agent in nutrient recycling.

For most land plants the soil is the only source of mineral nutrients and only roots in the soil are able to take up these nutrients. Again, the contrast with marine systems is striking. Plants in the sea absorb minerals from the waters that bathe them. Intertidal seaweeds suffer nutrient deficiencies during aerial exposures.

Plants take up minerals in an ionic form. In the soil ionic minerals occur in solution in soil water, and adsorbed to soil colloids. Only about 0.2% of the nutrient supply is dissolved in soil water. About 2% is adsorbed onto colloids, and the remainder consists of nutrients bound to organic detritus (humus), or incorporated in minerals or other insoluble compounds. Most of the nutrients are thus held in a reserve that is unavailable to plants. There is, however, a nutrient equilibrium between soil solution, soil colloids and reserves of minerals. As organic substances break down (through microbial activity) and minerals weather, the nutrients released are captured by the soil colloids. The binding is reversible so that the colloids act as ion exchangers. The colloids themselves are clay or humic substances. The ion exchange surfaces are in equilibrium with the soil solution. If ions are added to or removed from the soil solution (by plant roots) exchange takes place with the colloids.

The acidity of soils has a great effect on the ion exchange equilibrium and therefore influences plant nutrition. In very acid soils K^+, PO_4^{3-} and Mg^{2+} are soon depleted. Aluminum may accumulate to toxic levels. Alkaline soils are deficient in Fe, Mn and PO_4^{3-}.

One final problem associated with the shortage of available water in the atmospheric phase of terrestrial systems relates to sexual reproduction. Gamete fusion can occur only between cells surrounded by their cell membranes alone. Gametes cannot have cuticles or other protective coverings. In aquatic systems unprotected gametes are shed by all kinds of organisms and fertilization is usually external. An unprotected gamete would soon perish in the dry conditions of the atmosphere. There is in all successful terrestrial groups of organisms (seed plants, insects and vertebrates), a tendency toward internal fertilization and protection of the zygote.

The shortage of water in the earth's atmosphere is compounded by a very low CO_2 concentration. The CO_2 content of the atmosphere rarely exceeds 0.03-0.04% by volume and may limit the photosynthesis and growth of plants. In aquatic systems photosynthesis is rarely, if ever, carbon limited. In the sea CO_2 in solution comes from the atmosphere and from HCO_3^- and CO_3^{2-} ions in the water. As CO_2 is removed by photosynthesis it is replaced by the reaction of HCO_3^- ions with H^+ ions:

$$HCO_3^- + H^+ \rightarrow CO_2 + H_2O$$

The bicarbonate ions (HCO_3^-) are in equilibrium with the carbonate ions (CO_3^{2-}).

Plants on land have to deal with conflicting interests. On the one hand, they need to exchange as much gas as possible in order to obtain the CO_2 they need for growth. On the other hand, gas exchange will lead to loss of water vapor to the atmosphere. Land plants have a great capacity for drought avoidance, but their ability to tolerate reduced tissue water content is very limited. As will be shown in the following chapters, the adaptations of vascular plants to deal with these conflicting interests are biochemical, physiological, anatomical and morphological.

6
Mechanical Properties of Land Plants

The mechanical properties of terrestrial plants are to be found at all levels of organization. At the molecular level we must consider the attributes of **cellulose**. Cellulose is incorporated into the walls of cells and cells are arranged in distinctive mechanical tissues. At the organ level of organization we find distinctive mechanical roles for **prop roots, trunk buttresses** and so on.

Cellulose

The walls of plants cells are analogous to automobile tires. Embedded in an amorphous matrix of lignin and hemicelluloses are the microfibrils of cellulose. Using the car tire analogy, these are equivalent to the rubber and belts, respectively. Cellulose molecules have great strength (applied stress or force required before breakage occurs, measured as meganewtons/m^2). This is shown in Fig. 46. Cellulose has greater strength than steel, silk or nylon. The horizontal axis in Fig. 46 gives the **elastic modulus** of various materials. Elastic modulus (in

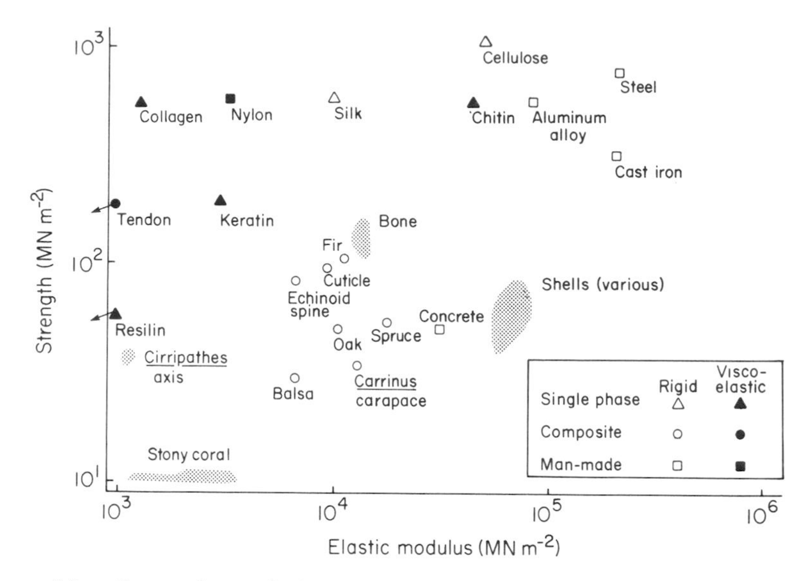

Figure 46. Strengths and elastic moduli of various natural and synthetic materials. Notice the superior properties of cellulose (modified after Wainwright, S. A. et al., *Mechanical Design in Organisms*. Copyright by Wainwright et al. 1976, 1982 by Princeton University Press. Reprinted with permission of Princeton University Press).

meganewtons/m^2) is stress (force per unit area) divided by strain (ratio of change in size or deformation). The elastic modulus of cellulose is among the highest of natural materials. The mechanical properties of cellulose can be attributed to its crystalline polymer structure. A crystalline substance is one whose atoms or molecules are fixed in a regular, three-dimensional array. In a linear polymer that is crystalline the molecular chains are packed in parallel arrays that are stabilized by a large number of attractive forces between neighboring atoms. Cellulose is highly crystalline giving a quite rigid structure. The elastic modulus of a highly crystalline polymer can be a thousand times greater than that of a comparable amorphous polymer. Crystalline cellulose is a linear polymer of D-glucose linked by *B* (1-4) glycoside bonds.

Crystalline cellulose is relatively inextensible when stressed by tensile forces in the fiber direction, but it is readily deformed by compression or forces normal to the fiber direction. Cellulose is the major mechanical component of plant cell walls, but cell walls do not behave as cellulose fibers do. The tensile strength of walls must be sacrificed for strength and rigidity in other directions.

The cellulose microfibrils of plant cells are wound helically. The helical fiber arrays permit shape and volume change of cylinders according to the fiber angle (the angle between the fiber and the long axis of the cylinder). The mean angle of all the fibers in a cell wall is called the preferred angle. The smaller the preferred angle, the greater the change in width of a cylinder per unit change in length. A low fiber angle gives great tensile strength, but requires little extension to break. Interestingly, the walls of rigid terrestrial plants have a preferred angle of 20° or less. In very elastic seaweed tissues the preferred angle is 60° which confers a high percentage extension to breaking point (~40%) without rupture of the cell walls.

Not all the cells of land plants have the same mechanical properties. Some cells with specialized mechanical functions are situated in special tissues. In general the rigidity of plants stems may be attributed to (a) turgidity of living parenchyma cells, (b) the presence of a mechanical tissue called **collenchyma**, and (c) woody tissue.

Turgor

The living cells of plants contain a solution of high osmotic potential so that there is a continuous hydrostatic water pressure pushing against the cell wall from the inside of the cell. The wall exerts a retaining pressure. The analogy to car tires can be used again here. The cell wall plays a role analogous to the tire itself and the hydrostatic pressure of cells plays the same role as air pressure in the tire.

The internal pressure of cells pushing against the retaining pressure of the walls produces turgor and imparts mechanical strength in plants. Loss of turgor occurs during drought. The visible result is wilting. Most vegetable crops collapse completely when there is insufficient water to maintain cellular turgidity.

The effect of cell turgidity is increased by the existence of tissue tension. In a dandelion stem the inner tissues are more extensible than the outer tissues. The outer tissues act as a resistance against which the inner tissues try to expand, so that in the intact stem the rigidity is increased. If the central tissues of a stem are isolated, they elongate while the outer tissues shorten (Fig.47), demonstrating the existence of tissue tension.

Collenchyma

Collenchyma cells form the chief tension-resisting tissue in wilting stems. Collenchyma cells are living, elongate structures (up to 2 mm long) that have thickened cell walls (Fig. 48). The cellulose thickening

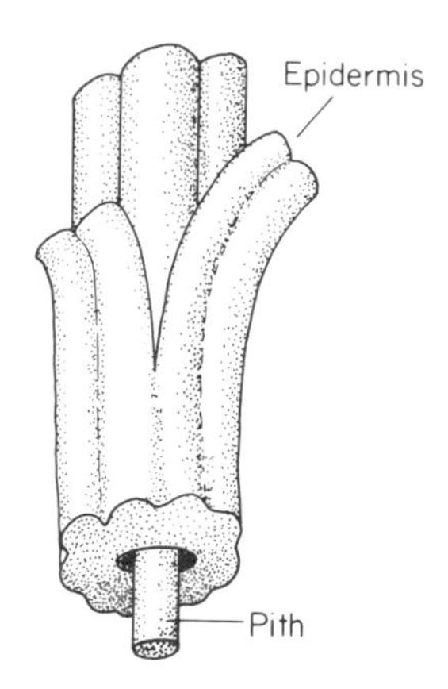

Figure 47. Effect of separating stem epidermis from underlying tissues with a corkborer. The epidermis shrinks and the pith elongates (modified from an original by K.E. von Maltzahn).

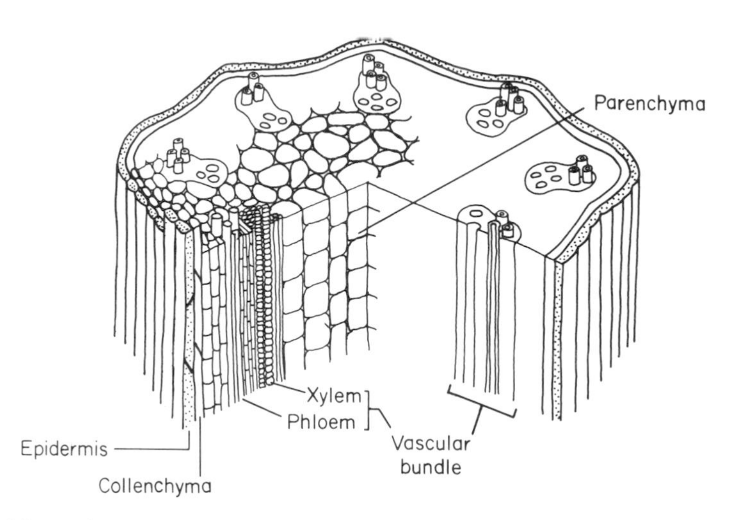

Figure 48. Three dimensional organization of tissues in a *Nasturtium* leaf stalk (modified after Wainwright, S.A. et al., *Mechanical Design in Organisms*. Copyright by Wainwright et al. 1976, 1982 by Princeton University Press. Reprinted with permission of Princeton Unversity Press).

in the walls is conspicuous at the corners of cells. In a cross-section of a stem the tangential walls of collenchyma cells are seen to be more thickened than the radial walls (Fig. 49). The tangential walls are fused to give a sandwich construction that has superior bend resistant properties. In the stems of non-woody plants collenchyma plays the major tension-resisting role, even in the presence of thickened

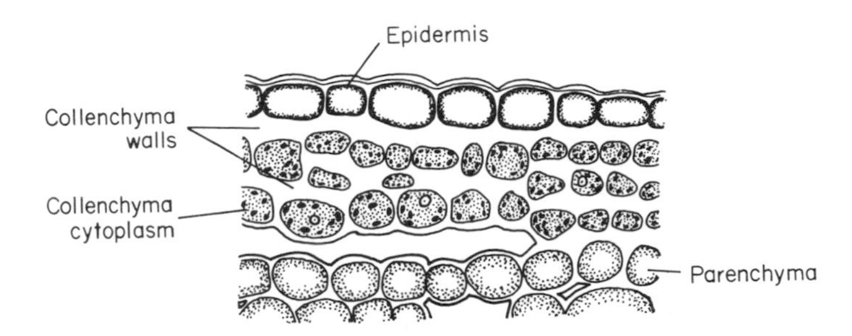

Figure 49. Cross section of the leaf of *Astrantia* to show thick collenchyma cells (modified after Haberlandt, 1914).

vascular tissues. Vascular tissues are inferior in the rigid resistance of tension. Because collenchyma cells are alive, they can grow as a plant elongates. For this reason collenchyma has an important mechanical function in growing organs.

Measurements on the strength of collenchyma have been made on dissected strands of tissue. It has been shown that collenchyma can support 10-12 kg/mm^2. When collenchyma strands are made to support only 1.5-2.0 kg/mm^2 they remain permanently extended. The tissue is therefore plastic and obviously suited for extension growth of young plants.

Within the stems of plants collenchyma is located close to the outermost cell layers and is therefore peripheral. The strengthened cells may form a continuous layer (Fig. 49) or may occur in strands that are visible as external ridges. Celery stalks fall into the latter category. The toughness of this vegetable can be attributed to extensive development of collenchyma strands.

Woody Tissues

There are four types of cells with mechanical functions in woody tissues. **Fibers** and **sclereids** have a mechanical role only. **Wood vessels** and **tracheids** conduct water through the plant, as well as performing mechanical functions. Woody tissues have secondarily thickened cell walls (that is, the thickening occurs after the cells are fully formed) and nearly all the cells are dead at maturity. The cells are important in supporting mature organs that have stopped extension growth.

Wood fibers and sclereids are often termed **sclerenchyma** tissue. Sclerenchyma cells are frequently scattered through the tissues of the plant body, but they are especially prominent in the vascular bundles

(Fig. 48). Fibers are narrow, elongated cells with sharp, tapering points. The walls are so thickened that only a narrow lumen remains (Fig. 50). Fibers usually occur in strands and these strands are of some commercial importance (hemp, jute, flax, linen, etc.).

Fibers are a little stronger than collenchyma cells and can support 15-20 kg/mm^2. A more interesting difference lies in the plasticity of the two cell types. Whereas collenchyma is permanently deformed by a weight of 1.5-2.0 kg/mm^2, fiber strands retain their original length after subjection to a tension of 15-20 kg/mm^2.

Sclereids are short, thick walled cells which are heavily **lignified** (lignin is a polymer of hydroxylated and methoxylated phenylpropane). The cells are frequently scattered among other cell types and may occur in stems, leaves, fruits and seeds. The seed coat characteristics of coconut give some idea of the mechanical strength of sclereids.

Vessels and tracheids are called the **tracheary elements** of wood. Their primary function is in the conduction of water from the roots to all parts of the plant in the atmosphere. Tracheids are dead, hollow cells that are elongated (0.5-3 mm long and $\sim 30 \mu m$ wide). The ends of the cells are narrow (Fig. 51) and the walls are heavily thickened. Water flows through the pits in the walls. The thick rigid walls play a role in the mechanical support of the plant.

Tracheids are thought to have given rise (in an evolutionary sense) to both fibers and wood vessels. Fibers appear to have arisen by an increase in wall thickness, decrease in length and reduction in the size of pits. Wood vessels evolved from tracheids (that were long, narrow and with tapering ends) by a process giving rise to elements that are short, wide and with slightly inclined transverse end walls.

The evolution of vessels and fibers may be regarded as a division of labor. The vessels are concerned with water transport and the fibers with mechanical support. Tracheids must do both jobs. The structure and mechanics of tracheid walls have been the subject of intensive investigation. Tracheid cells are bonded together in a matrix of amorphous materials. The walls of each tracheid are laminated and each lamina has a different preferred microfibril orientation (Fig. 52). The cellulose microfibrils are the primary load-bearing components of the tracheid wall. Xylans (polymers of the sugar xylose) act as an intermicrofibrillar glue. Lignin acts a bulking agent that controls the hydrophilic properties of the wall. The tight crosslinking of the microfibrils in the tracheid gives the cell a linearly elastic behavior similar to that of continuous crystalline cellulose.

Wood vessels are fused rows of wide, elongated cells (Fig. 53).

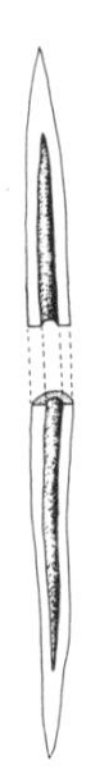

Figure 50. Longitudinal section of a xylem fiber (modified after an original by K.E. von Maltzahn).

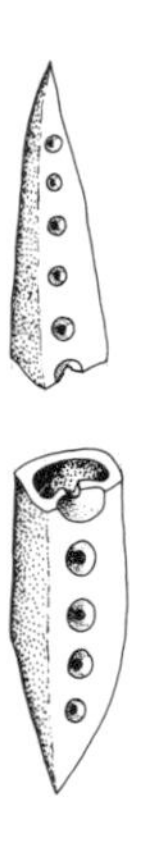

Figure 51. Xylem tracheid cell with thickened walls perforated by pits for water transport (modified after an original by K.E. von Maltzahn).

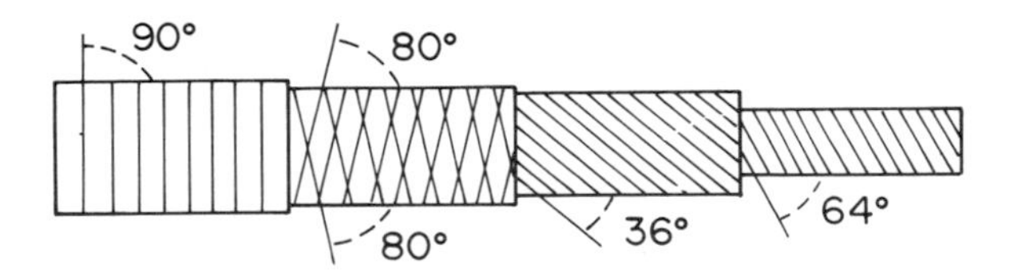

Figure 52. Orientation of cellulose microfibrils in successive laminations of a tracheid cell wall. The oldest lamination is on the left and the newest on the right (modified after Mark, 1967).

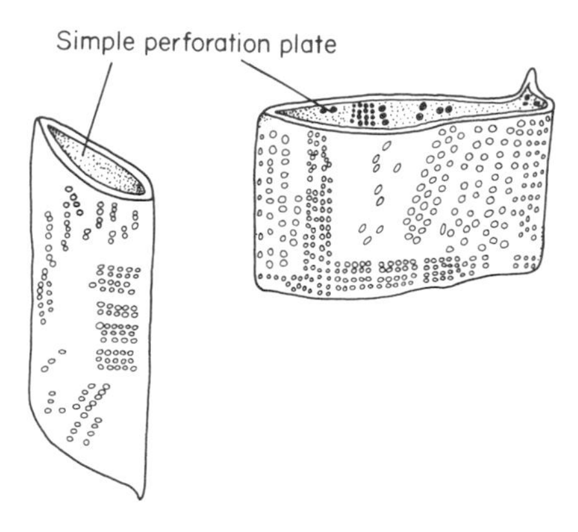

Figure 53. Xylem vessel cells from oak. Each cell is part of a stacked series joined end to end at the perforation plates (modified after Esau, 1976 © by John Wiley & Sons, Inc.).

The walls of vessels usually have spiral or annular thickenings like those which give rigidity to vacuum cleaner hoses. Vessels have less of a mechanical role than tracheids, and it is thought that wall thickenings of vessels are designed to resist implosive forces placed on them during water conduction.

Arrangement of Mechanical Tissues

Collenchyma and woody tissues are not uniformly distributed in the organs of plants. Furthermore, the mechanical elements are differentially distributed in stems, roots and leaves. This is a reflection of the different strains to which these organs are subjected. Cylindrical stems are subjected to bending forces acting in any plane at right angles to the longitudinal axis. Flat organs, like leaves, are chiefly exposed to bending stresses in a plane which is perpendicular to their greatest surface. Roots are subjected to linear tension and compression, but not to bending forces. When a stem bends the stress is distributed peripherally (Fig. 54). It is therefore easy to explain the peripheral distribution of mechanical elements in non-woody plants. Engineers achieve an efficient distribution of strengthening material to deal with bending through the use of the familiar I-beam girder. I-beam construction is not common in plant stems. The reason is that bending of stems may occur in several planes, while an I-beam resists bending in only one plane. Generally we see box-sections or tubular-sections in stems (Fig. 55).

The primary growth of vascular plants is localized in apical

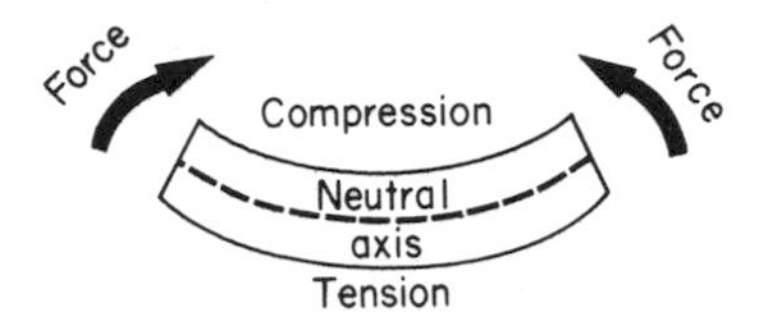

Figure 54. Bending of a linearly elastic beam in the direction of the arrows, causing compression on one surface and tension on the other (modified after Wainwright, S. A., et al., *Mechanical Design in Organisms*. Copyright © by Wainwright, et al. 1976, 1982 by Princeton University Press. Reprinted with permission of Princeton University Press).

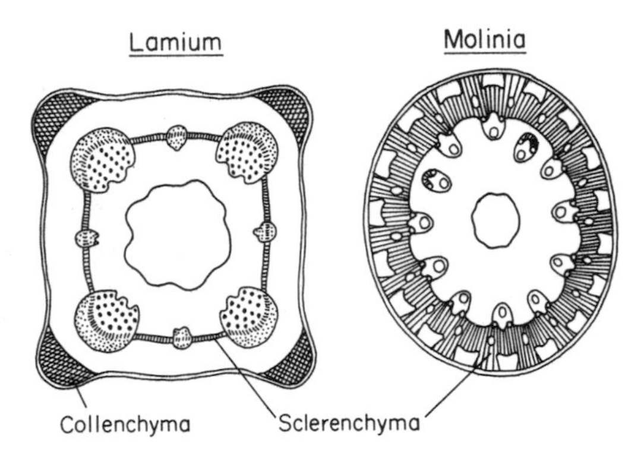

Figure 55. Arrangement of mechanical tissues (shaded) in two stems (modified after Skene, 1924).

meristems. The apical meristems of seedlings are relatively small, but they become progressively larger with the plant. The progressive enlargement of the bud produces a primary stem which tapers inward toward the soil. The result of this is a top-heavy structure. Some plants deal with the top heavy distribution of mass by growing buttress roots or stilt roots (e.g., mangroves or corn, Fig. 56). These aerial roots are subjected to compression and extension stress and have extensive peripheral fiber development. Most plants have **secondary thickening** mechanisms to deal with an increasingly top-heavy construction. For a stem that is fixed in the soil and bearing a weight of foliage, the bending moment is greatest at ground level. It is therefore essential to have a concentration of mechanical elements in the lower stem. Plants that increase in circumference through secondary thickenings develop a cylinder of dividing cells within the stem. This meristematic cylinder is called a **cambium** (Fig. 57). Most of the tissues produced by the cambium are formed from cells on its inner face and consist of tracheids, vessels and sclerenchyma fibers. In trees

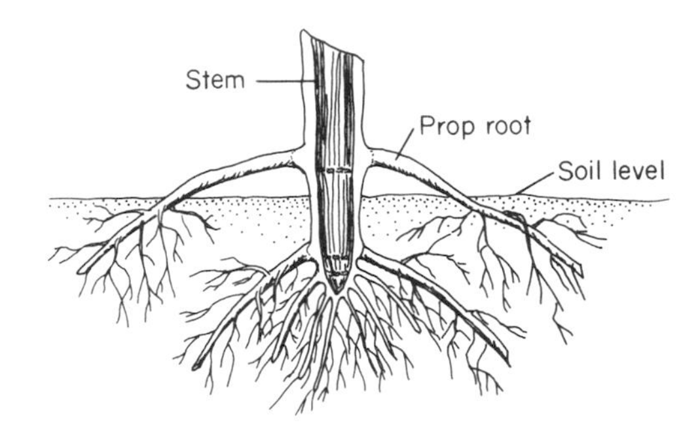

Figure 56. Prop roots of corn (modified after Haberlandt, 1914).

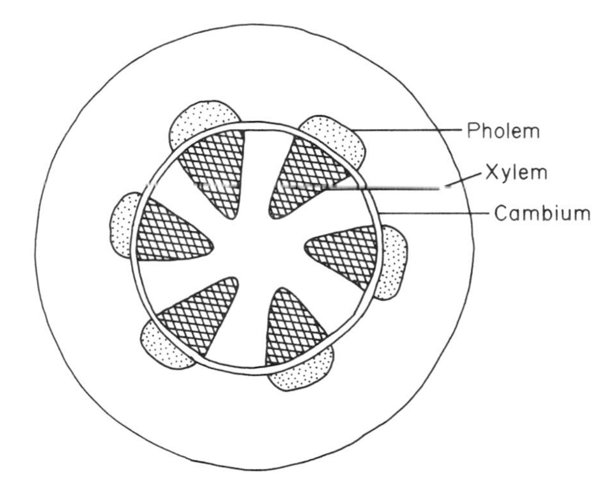

Figure 57. Cross section of a stem showing the formation of a meristematic cambium cylinder which produces secondary thickening (modified after an original by K.E. von Maltzahn).

this secondary agglomerate is commonly called wood, and it comes to occupy the bulk of the axis with no special differential arrangement.

Wood is structurally complex in four ways relating to various levels of organization. At the molecular level it comprises crystalline cellulose in an amorphous matrix of lignin, hemicelluloses and other organic compounds. Above the molecular level wood is an aggregate of cell walls arranged in cylinders that lie parallel to the long axis of the stem. Cellulose microfibrils have a preferred orientation that varies with its age in the wall (Fig. 52). Finally, wood of many species occurs in concentric annual growth layers with varying ratios of cell wall:cell lumen dimensions. This combination of attributes makes wood a very complex agglomerate (much more complex than, for instance, reinforced concrete) and little is known about its mechanical properties in live trees. Much of the information available on the properties of dried timber is not strictly relevant.

Roots in soil are subjected to tension and the resistance to such strain is directly proportional to the area of the cross-section. The arrangement of mechanical tissues in the cross-section of a root is not relevant to the resistance of tension. Generally the mechanical tissues of roots or underground stems are concentrated in an axile bundle (Fig. 58) which contrasts with the peripheral distribution in herbaceous stems (cf. Fig. 55).

Leaves are often large flat structures that are generally subject to bending forces in one plane only. It is therefore in leaves that we see the best examples of I-beam construction with all the girders arranged parallel with one another and at right angles to the leaf surface (Fig. 59).

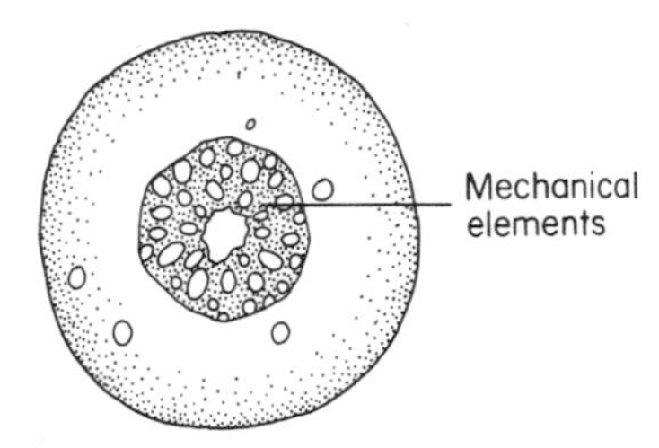

Figure 58. Cross section of an underground stem of *Carex* showing the axile distribution of mechanical elements (modified after Haberlandt, 1914).

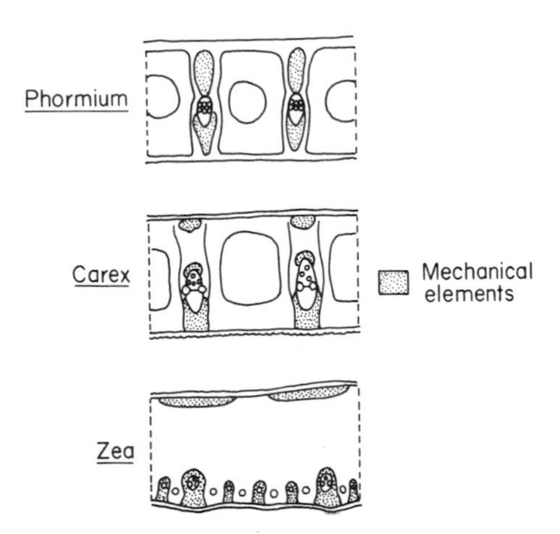

Figure 59. Cross sections of three leaves showing distribution of mechanical elements in parallel girder formations (modified after Haberlandt, 1914).

The development of mechanical tissues is often a function of wind conditions during growth. When corn grown for 40 days under moderate gale conditions is compared with plants grown in still air, considerable differences in anatomy are revealed. Gale grown leaves are twice as thick as control leaves. The margins of exposed leaves have 12 rows of sclerenchyma cells as opposed to 2 rows in controls. There are about three times as many fibers in the vascular bundles of exposed plants. In experiments with celery, wind motion can stimulate a 50% increase in collenchyma development. In addition, the growth in height of cereals is reduced under windy conditions. Obviously plants are very sensitive to mechanical stress and adapt their anatomies and morphologies during development to meet prevailing conditions.

Experiments with trees in wind tunnels have shown that, as wind velocity increases, the foliage crown compacts to present a reduced cross-sectional area. The net effect is that pressure drag increases linearly with wind velocity, whereas for rigid objects the drag is proportional to the square of the velocity. This behavior is exactly paralleled in foliose seaweeds subjected to surge force from oscillating waves. In other respects the mechanical adaptations of land plants at the cell, tissue and organ levels of organization are all quite distinctive.

7

Water Absorption, Water Loss, Carbon Dioxide Exchange and Vascular Transport

The topics in the title of this chapter seem, at first, quite disparate. For land plants the processes listed are interconnected in such a way that they cannot be considered separately.

With adequate illumination the photosynthetic systems of most land plants utilize CO_2 very inefficiently at ambient atmospheric concentrations. This inefficiency is directly attributable to the activity of the carboxylation enzyme **ribulose biphosphate carboxylase**. This enzyme evolved in aquatic plants which do not suffer from a CO_2 deficiency. In the atmosphere the enzyme works at only one-fifth to one-third of its potential (efficiency is measured as the number of CO_2 molecules fixed per molecule of enzyme). One way of dealing with CO_2 starvation is to increase the concentration of carboxylase in the chloroplasts. However, ribulose biphosphate carboxylase is a very large molecule with a low turnover rate and is already present in such high concentrations that it sometimes exists as paracrystalline arrays rather than in true solution. The enzyme loses even more of its efficiency when in these arrays. Therefore plants on land have to exchange large quantities of gas in order to obtain enough CO_2 for photosynthesis. Unfortunately no material has yet been produced that is highly permeable to CO_2, but impermeable to water vapor. Since the water concentration in air is less than that in equilibrium with the wet walls of photosynthetic cells, acquisition of CO_2 involves enormous water losses through open pores.

Water Movement in Plants

Water moves upward through a plant as a result of the water potential gradient (see Chapter 5) between atmosphere and soil. For water to move it is necessary that $\Psi_{soil} > \Psi_{root} > \Psi_{stem} > \Psi_{leaf} > \Psi_{air.}$ In each part of the system there will be a resistance to flow. A summary of water potentials and resistances is shown in Fig. 60. The actual flow rate depends on the following relationships:

$$\text{flow rate} = \Psi_{root} - \Psi_{leaf} / R_{root} + R_{stem} + R_{leaf}$$

where R is the resistance to flow. Therefore the rate of flow is proportional to the water potential difference and inversely proportional to the resistances.

The key to the movement of water through the plant is the great capacity that dry air has for water vapor. As the relative humidity (RH) of air drops below 100%, its affinity for water increases dramatically. At 100% RH the water potential (Ψ) of air is zero. At 20°C and a relative humidity of 98% the water potential drops to -27.2 bars, which is sufficient to hold a column of water 277 m high. At 50% RH the water potential of air is -935 bars. The significance of these very low water potentials is as follows. The soil solution rarely has a water potential of less than -15 bars, so that air does not have to be very dry in order to establish a steep water potential gradient from the soil, through the plant and into the atmosphere The driving force in the gas phase is 10-100 times higher than the driving force in the liquid phase. Within the plant the lowest resistance to water movement (highest conductivity) occurs within the tracheary elements (tracheids and vessels). The conductivity of tracheids is one million times higher than that of ordinary parenchyma cells. Thus it is hardly surprising that tracheary elements are the major pathways of long distance liquid transport in plants. However, the conducting

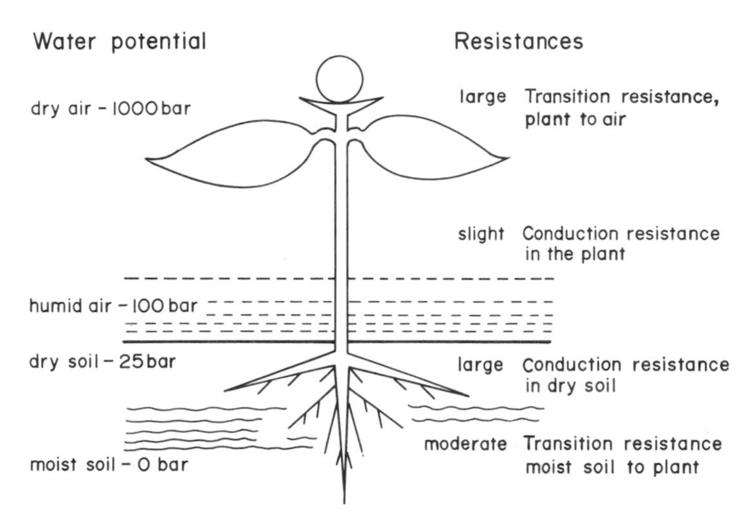

Figure 60. Gradients of water potentials and resistances between soil, plant and atmosphere. The water potentials are order of magnitude estimates. Notice the very low water potential of dry air. Notice also that humid air has a lower water potential than dry soil and that this gradient drives the flow of water. The highest resistance to water flow is in the transition between leaf and air (modified after Larcher, 1975).

tissues of roots have an axile distribution (see Chapter 6), so that water absorbed by root hairs must travel through several layers of parenchyma cells before reaching the conducting tissues. Similarly, in the leaves, water has to be transported through several cell layers from the tracheary elements to the evaporating surfaces of cells in contact with air. Apparently, the total resistance to water flow in these short parenchymatous pathways in non-woody flowering plants exceeds the overall resistance of the much longer tracheary element pathways.

Apart from the water potential gradient and resistances to flow, the ascent of sap is dependent on the **cohesion of water**. This is the mutual force of attraction between water molecules in the transport pathway. This force of attraction is so great that when water is pulled by osmosis and evaporation at the top of a tree, the pull extends all

Regulation of Water Loss

Land plants are **homiohydric**, that is, they regulate their tissue water content under varying conditions of water supply and loss. Algae are **poikilohydric** and their water content fluctuates according to environmental conditions.

The primary structures responsible for the homiohydry of land plants are the **cuticle** and the **stomata**. The cuticle prevents water loss and the stomata regulate water loss (and CO_2 acquisition).

The cuticle is a water-resistant layer formed on the outermost surfaces of plant cells in contact with the atmosphere. The cuticle is, of course, absent from gas exchange and water-absorbing surfaces. The water-resistant surfaces of cells in atmospheric contact are complex (Fig. 61). The cuticle itself is a varnish-like layer of polymerized fatty acids and soaps which are synthesized in the cytoplasm. The cuticle is covered with wax on its outermost surface. Inside the cuticle is the wall which may contain wax plates that slow water movement.

The cuticle is very resistant to water loss. In cacti the loss by evaporation is only 0.05% of the loss from a free water surface. At most, cuticular loss of water is only about 30% of that which is lost from the plant as a whole. The usual cuticular loss is 5-15% of total plant loss. Algal thalli have no structures that are functionally equivalent to the cuticle.

The cuticle is more resistant to CO_2 movement than it is to water vapor movement. Therefore CO_2 exchange occurs on cuticle-free cell surfaces and these are to be found in small subsurface air chambers of the leaf which are connected to the atmosphere through minute pores (Fig. 62). Gas exchange occurs on internal surfaces, as it does in all successful terrestrial groups of organisms (insects, tetrapod vertebrates, as well as vascular plants). The size and spacing of the pores that link the internal air chambers with the atmosphere are vital determinants of the efficiency of CO_2 exchange. The pores usually occupy only 2% of a leaf's surface area. You might think that this would reduce CO_2 exchange to only 2% of a free absorbing surface of

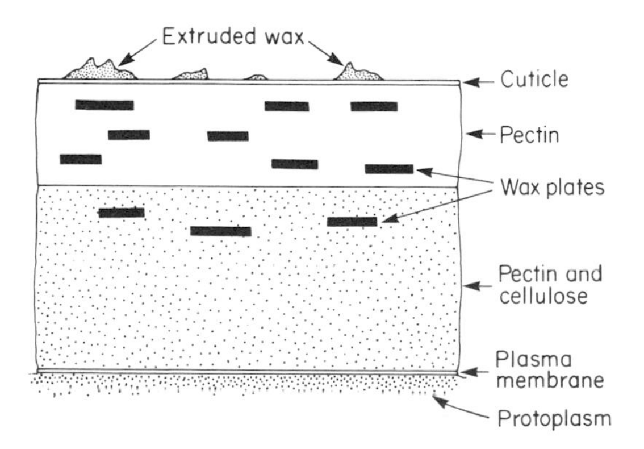

Figure 61. Section of an epidermal cell wall showing cuticle and extruded waxes (modified after Leopold and Kriedemann, 1975).

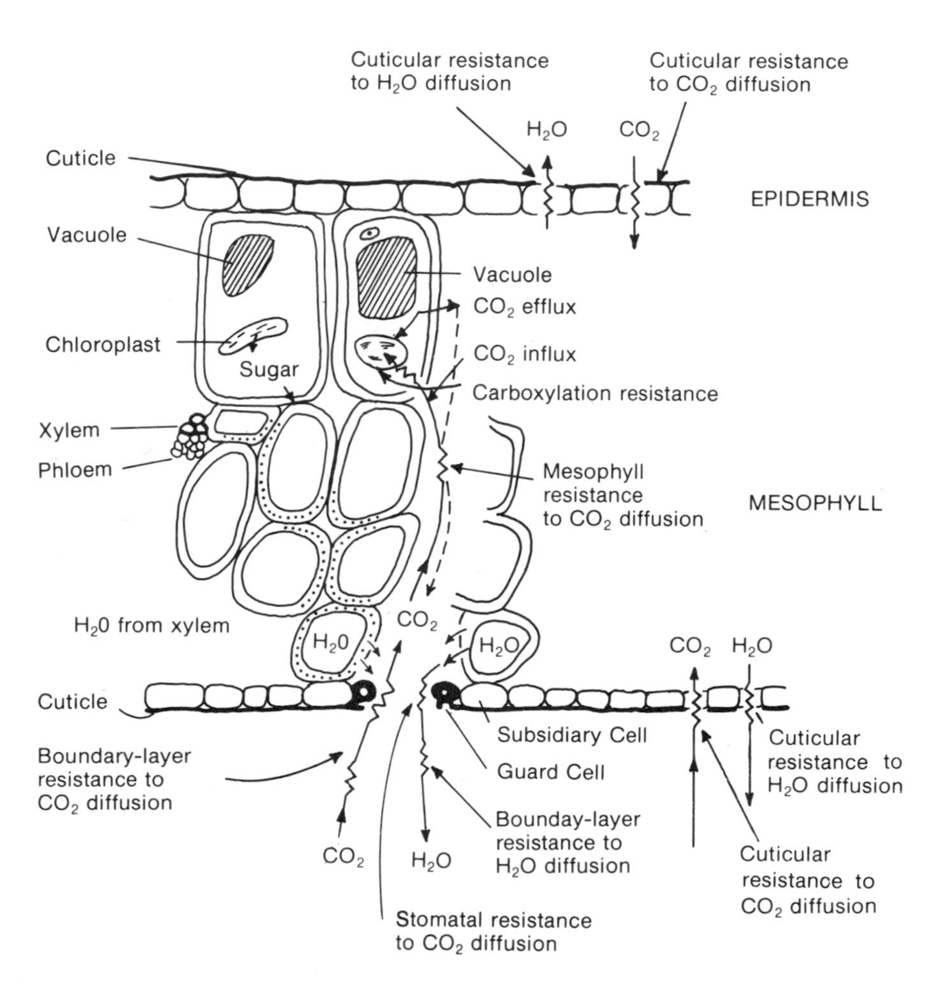

Figure 62. Section of a leaf showing substomatal air chambers, guard cells, pathways of H_2O and CO_2 movement and resistances to gas exchange (modified after Leopold and Kriedemann, 1975. *Plant Growth and Development.* © 1964 McGraw-Hill Book Company. By permission).

the same area as the leaf. This is not so. A leaf can absorb CO_2 at 25% of the rate of a free absorbing surface of sodium hydroxide.

Diffusion of gas through small pores is much different from the diffusion of gas from a free surface. The rate of diffusion through small pores is proportional to the diameter (and perimeter) of each pore, but not to its area. The diameter of a pore is proportional to the square root of the area. This means that if you have a pore that is half of the area of another pore, gas diffusion through the smaller pore will be 70% of the diffusion through the larger pore (not 50%). Thus the smaller a pore, the greater is diffusion per unit area. These relationships apply both to water vapor and CO_2 diffusion.

The reason why gases diffuse faster through small pores can be explained by considering gas concentration gradients (Fig. 63). If a perforated membrane is placed over a dish of water, the atmosphere below the membrane becomes saturated with water vapor and the gas diffuses out through the perforations. Over each pore there is a hemispherical dome of decreasing water vapor concentration (Fig. 63). The larger the pore diameter, the larger the hemispherical shell. Water vapor molecules diffuse faster down a steep concentration gradient than they do down a less steep gradient. This means that gas molecules move faster down the steeper concentration gradient of the diffusion shell of small pores than down the less steep gradient of the large diffusion shells of larger pores. For this system to work, it is essential that the pores be widely spaced enough to prevent a single large shell forming over a number of small pores (Fig. 63). Hence the size and spacing of pores are essential design considerations for gas exchange.

The small pore situation in Fig. 63 (diffusion shell on one side of membrane only) applies especially to the movement of water vapor from the saturated air chambers of the leaf. Carbon dioxide forms concentration gradient shells on the inside and on the outside of the pore (while there is photosynthetic CO_2 consumption). The net effect is that CO_2 molecules move into a leaf faster than water vapor molecules move out.

The regulation of water loss in vascular plants involves a variable resistance to gas exchange. The resistance is varied by opening and closing the pores of the internal air chambers. When soil water supply is reduced the plant responds by closing the air chamber pores. This reduces water loss greatly, but also restricts the entry of CO_2 into the plant. The variable aperture pores are in the gas diffusion part of the water transport system. This has the greatest resistance (lowest conductivity) to water flow of any part of the water-flux route. This means that a small change in resistance has a much

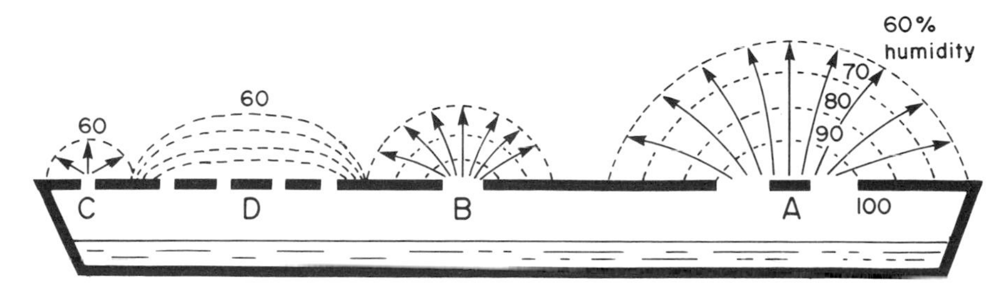

Figure 63. Gas concentration shells (shown as % humidity contours) over perforations in a membrane covering water-saturated air (modified after Coult, 1973 by permission of Longman Group Ltd.).

greater effect on water movement (and loss) than a similar change in the liquid phase of the system.

The variable aperture pores that regulate gas exchange are enclosed by **guard cells** (Fig. 62). The guard cells change their shape to open and close the pores. The pores and guard cells together form units known as **stomata**. Guard cells change shape when their turgor changes. If the cells become turgid, they change shape in such a way that they open the pores. If the cells are flaccid, they close the pores. The turgidity of cells depends on their osmotic potential. When this is high, water pressure inflates the cells. The osmotic potential is controlled by pumping potassium ions into and out of the guard cells. Water stress leads to the production of **abscissic acid** (an organic acid) which drives K^+ ions out of the cell causing the stomatal closure. Stomata are also sensitive to CO_2 concentration. Lowered CO_2 tension in the substomatal chambers leads to an opening of the pores. There is, however, an overriding control which closes the stomata when water supply is low. It is interesting in this regard to note that the osmotic potential of open guard cells (about -15 bars) is the same as the total soil moisture level commonly regarded as the wilting point for plants.

The substomatal air chambers are an essential component of the homiohydric gas exchange system. Stomata opening into a tissue without air spaces would not provide an effective CO_2 supply to chloroplasts more than $10 \mu m$ from the pore. The photosynthetic rate at light saturation would be reduced to 10% of that which occurs in plants with air chambers.

In 1984 it was reported that there is one vascular plant that has solved its CO_2 supply problem in a unique way. In the high Andes of Peru there is a rare fern called *Stylites andicola* which has no stomata, but has a thick cuticle that is resistant to CO_2 exchange. The roots of *Stylites* absorb CO_2 from the soil water solution where mineralization of organic plant remains ensures an elevated inorganic carbon supply. The CO_2 absorbed by the roots is transported to the green above-ground parts of the plant where carbon reduction takes place.

Metabolic Adaptations to Water Loss and Carbon Dioxide Starvation

Some groups of flowering plants have evolved new metabolic pathways that deal with the serious problem of low CO_2 affinity characteristics of ribulose biphosphate (RuBP) carboxylase. A new carboxylation enzyme (**phosphoenol pyruvate** [PEP] **carboxylase**) has evolved.

This new enzyme has a much greater affinity for CO_2 than RuBP carboxylase. PEP carboxylase endows plants with a high efficiency for CO_2 uptake under reduced water conditions. The reason for this is that net photosynthesis can be maintained at much lower substomatal CO_2 concentrations than in plants that lack the enzyme. This means that gas exchange (and water loss) can be reduced while a positive rate of photosynthesis is maintained. One group of flowering plants that has PEP carboxylase comprises the drought-resistant cacti and other desert succulents. These plants are different from all others in that their stomata are closed during the day (when photosynthesis is taking place) and open at night. This may seem aberrant behavior in view of the fact that CO_2 is needed for photosynthesis to proceed. In fact these succulents store CO_2 in the dark. Carboxylation by PEP carboxylase in the dark leads to an accumulation, of organic acids inside the cell vacuoles (Fig. 64c). In the light, these acids are decarboxylated, releasing CO_2 which is immediately combined with ribulose biphosphate (catalyzed by ribulose biphosphate carboxylase). This type of photosynthesis is called **crassulacean acid metabolism** (CAM). The advantage of the CAM mechanism is obvious. Evaporative losses are least during the night. Thus, CAM plants reduce water loss while maintaining net photosynthesis during the day.

In another group of flowering plants that has PEP carboxylase the stomata are open during the day. Carbon dioxide is assimilated into organic acids in the cells surrounding the air chambers of the leaves. The organic acids are then transported to a dense layer of large, thick-walled, chloroplast containing cells that form a sheath around the vasular bundles of the leaf. Accumulated organic acids in the bundle sheath cells are decarboxylated, releasing CO_2 which is incorporated into the photosynthetic Calvin cycle. This leads to the

Figure 64. (a) C_3 dark reactions of photosynthesis in a leaf mesophyll cell. Incorporation of CO_2 into the Calvin cycle involves combination with ribulose biphosphate [RuBP] (catalyzed inefficiently by RuBP carboxylase). (b) C_4 dark reactions in mesophyll and bundle sheath cells. CO_2 is combined with phosphoenol pyruvate [PEP](catalyzed efficiently by PEP carboxylase). Malic acid and aspartic acid are transported to bundle sheath cells and decarboxylated to produce CO_2 that is incorporated into the Calvin cycle. (c) CAM dark reactions. CO_2 is combined with PEP to form malic and isocitric acids at night. The acids are decarboxylated during the day to supply the Calvin cycle (modified after Barbour et al., 1980).

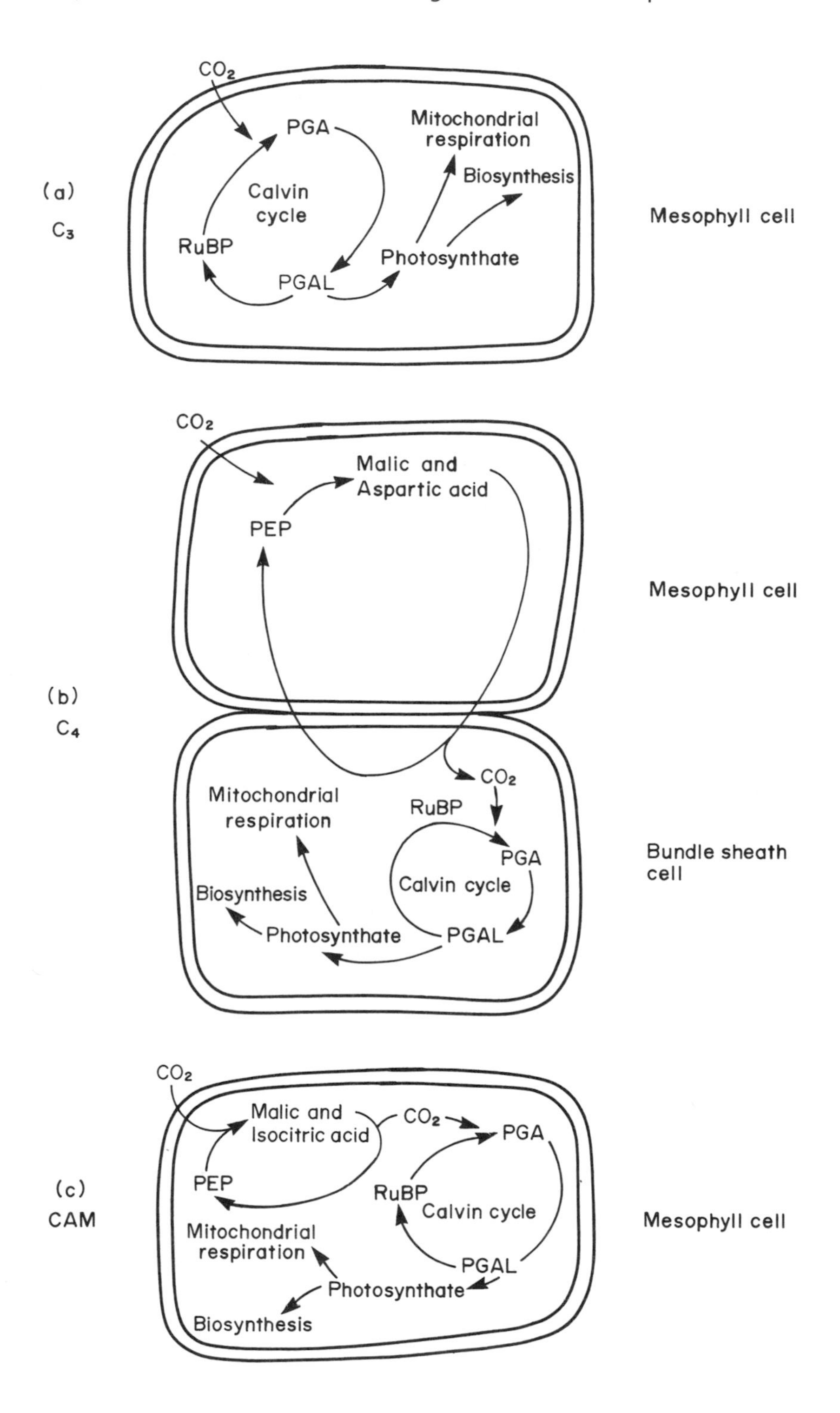
(a)
C_3
CO_2
PGA
Calvin cycle
RuBP
PGAL
Mitochondrial respiration
Biosynthesis
Photosynthate
Mesophyll cell
(b)
C_4
CO_2
Malic and Aspartic acid
PEP
Mesophyll cell
CO_2
Mitochondrial respiration
RuBP
PGA
Calvin cycle
Biosynthesis
Photosynthate
PGAL
Bundle sheath cell
(c)
CAM
CO_2
Malic and Isocitric acid
CO_2
PGA
PEP
RuBP
Calvin cycle
Mitochondrial respiration
PGAL
Photosynthate
Biosynthesis
Mesophyll cell

formation of hexose (Fig. 64). Plants with this type of anatomy and photosynthetic carboxylation machinery are called **C_4** plants (the organic acids produced are 4 carbon structures). Plants that do not have PEP carboxylase are called **C_3** plants. The advantages of the C_4 metabolism stem from the water use efficiency of PEP carboxylase and from the fact that CO_2 can be concentrated in the bundle sheath cells. RuBP carboxylase is most efficient at high CO_2 concentrations.

Transport of Minerals

The water transport route of land plants also serves as the pathway for mineral nutrients from the soil to the aerial organs. A root hair takes nutrients from the soil in the following ways (Fig. 65):

1. Absorption from the soil solution (ions are pumped through the root hair membrane into the cytoplasm),

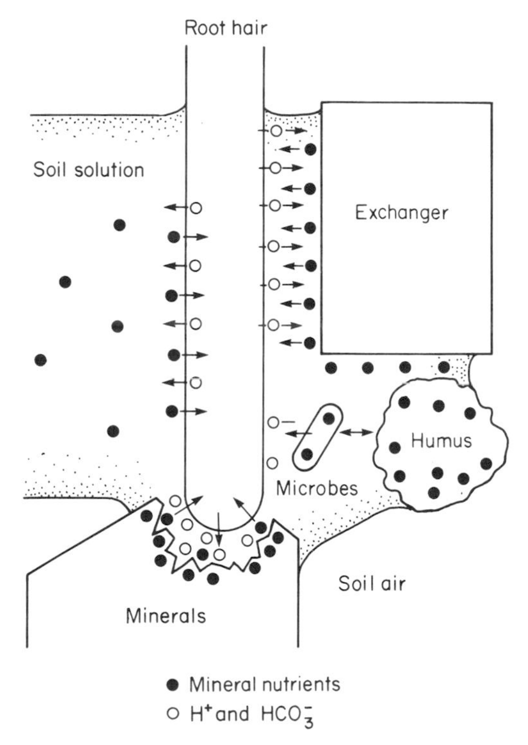

Figure 65. Mobilization and uptake of mineral nutrients by roots in the soil. The exchanger is a soil colloid which releases nutrient ions in exchange for H^+ and HCO_3^- ions produced by the root hair. Ions are also taken from the soil solution directly. H^+ and organic acids secreted by the root hair liberate chelated mineral complexes (modified after Larcher, 1975).

2. Ion exchange with soil colloids (H^+ and HCO_3^- are released by the root hair and exchanged for adsorbed nutrient ions),
3. Freeing bound nutrients (H^+ and organic acids excreted from the roots liberate chelated mineral complexes which are taken up by the roots).

It should be pointed out that land plants have a major disposal problem with excess H^+ or OH^- ions generated primarily in relation to the assimilation of ammonium and nitrate. Therefore the excretion of H^+ is a requirement for ion exchange and for waste disposal. Nitrogen assimilation often occurs in underground portions of the plant and so there is a ready supply of ions available for exchange. When assimilation of nitrate occurs in aerial shoots, the excess OH^- is neutralized in organic acids. The acids are transported to the root where they may be excreted directly, or first decarboxylated, thus regenerating OH^- ions.

Once in the root hairs ions travel through water imbibed cell walls or through the cytoplasm of living cells to the vascular tissues. Most minerals are transported to mature photosynthetic tissues because this is the route of water transport. However, mature tissues need little mineral nutrition in comparison with the apical meristems. Transport from the photosynthetic tissues to the growing parts occurs through a different kind of conducting element, the **phloem tissue**.

Transport of Organic Solutes

Phloem tissues are mainly responsible for the transport of organic molecules such as sugars and amino acids. Of the cell types in the phloem, the **sieve elements** play the major role in translocation. The function of these cells is analogous to that of the sieve filaments of brown algae (see Chapter 4).

Phloem tissues run approximately parallel and peripheral to the xylem (Fig. 48). Thus all of the conducting elements of plants connect roots in the soil with the organs in the atmosphere (stem, leaf, reproductive structures). Again, it must be emphasized that this conducting system is essential because each vascular plant inhabits two different worlds, soil and atmosphere.

There are two kinds of sieve elements, **sieve cells** and **sieve tube members**. Sieve tube members are found only in flowering plants. Other vascular plants have sieve cells only. A sieve cell resembles a tracheid in that it is an elongated structure, tapered at both ends.

Translocation occurs trough sieve areas that perforate the sieve cell walls (Fig. 66). Sieve cells are, however, alive and thin walled in comparison with tracheids.

Sieve tube members (Fig. 67) are specialized structures which are

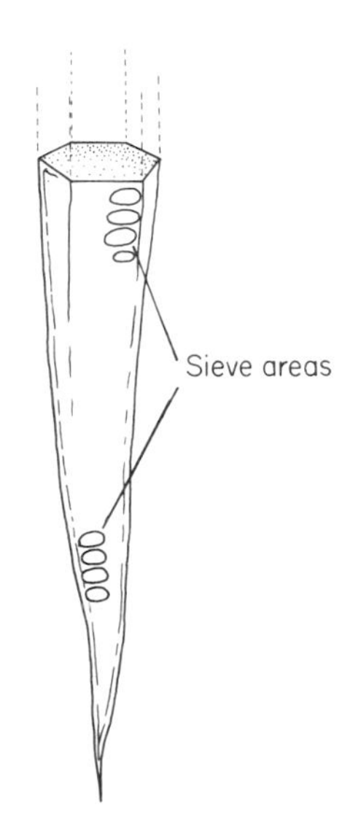

Figure 66. A fern sieve cell showing sieve areas (modified from an original by K.E. von Maltzahn).

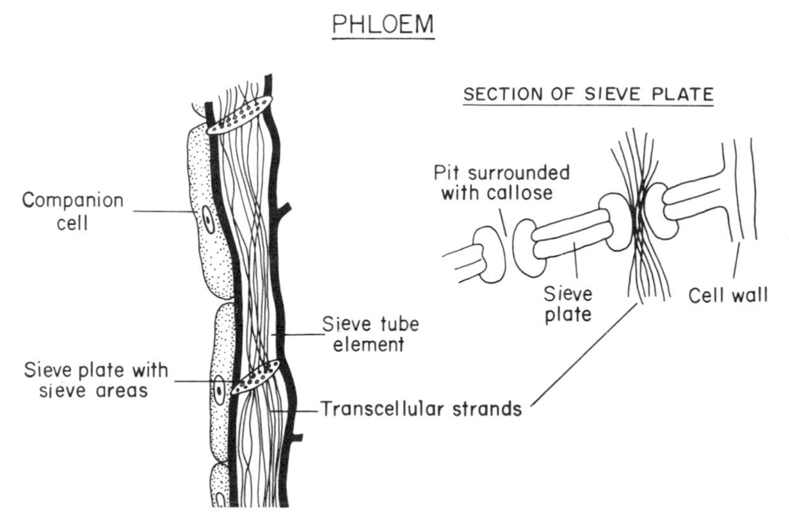

Figure 67. Sieve tube elements and companion cells (modified after Leopold and Kriedemann, 1975. *Plant Growth and Development.* © 1964 McGraw-Hill Book Company. By permission).

arranged in longitudinal pipe-like arrays. The nucleus of a differentiated sieve tube disintegrates and so does the membrane surrounding the cell vacuole. Cytoplasmic strands run the length of each sieve tube and pass from cell to cell through the sieve plates. The strands contain microfibrillar arrays.

The enucleate sieve tube members are closely associated with companion cells (Fig. 67) that have prominent nuclei. The companion cells have direct cytoplasmic continuity with the sieve tubes and probably provide for the metabolic requirements of the tubes.

After at least 50 years of study there is no generally agreed mechanism to account for the flow of solutes through phloem cells. The best known theory (the Münch pressure flow hypothesis) postulates a physical mechanism that is independent of metabolic energy. However, phloem transport is a highly energy-dependent system. Because of uncertainties about the phloem transport mechanism, the topic is not pursued further here.

Cell Structures of Vascular Plants in Relation to Problems of Life on Land

Chapters 6 and 7 introduced special types of cell structures that have evolved in the vegetative bodies of vascular plants to deal with the problems of terrestrial existence. These can be summarized in relation to the problems examined in Chapter 5.

Mechanical support: The following types of cells have evolved in relation to the problems of mechanical support in a non-buoyant medium: **collenchyma**, **sclereids**, **fibers**, and **tracheids** (vessels probably do not have a mechanical support role). The mechanical cell types are distributed within the plant body so that they counteract the forces of **bending**, **tension** and **compression**. None of the cell types listed here occurs in algae.

Drought and problems associated with water supply: The following cell structures have evolved in relation to water supply and CO_2 availability: **guard cells**, **cuticle**, **root hairs**, **vessels** and **tracheids**. We may also include substomatal air spaces as structures related to gas exchange. Sieve element analogs occur in algae, but none of the other cell structures listed above is to be found among the seaweeds.

8

Drought Avoidance in Desert Plants

All vascular plants have morphological and anatomical features that make them homiohydric. The general features of drought avoidance were introduced in Chapter 6. Plants that live in deserts have special characteristics for dealing with extreme water deficiency stress. Most of these characteristics are morphological and anatomical in nature and they will be considered in this chapter. The special metabolism of some desert plants (CAM photosynthesis) has already been introduced.

Most of the stresses of life on land (e.g., temperature extremes) are tolerated or endured by plants. However, drought is avoided so that the tissue water content is regulated, even under desert conditions. There are two strategies by which water content is maintained under arid conditions: (a) **water retention** and (b) **accelerated water absorption**. Exponents of these strategies are called, respectively, **water savers** and **water spenders**. There is a third strategy for dealing with drought, this is the **ephemeral** strategy. Ephemerals are not drought avoiders, rather they are drought escapers. This group of plants survives periods of water stress as drought resistant seeds. The plant bodies are not well adapted to arid conditions. The seeds have resistant coats and a very reduced metabolic rate.

Water Savers

Plant morphologists have long been fascinated by the adaptive structures of desert plants. Long lists of structural features thought to be effective in retaining water were drawn up. Such structures were called **xeromorphic**. Subsequent experimental analysis has shown that many supposed xeromorphic structures lose water faster than parts of plants from moist locations. Nevertheless, true water saver plants do reduce water loss to a far greater degree than other plant groups. Water spenders may lose water half a million times faster than water savers. The adaptations of water savers to achieve this ability are to be found in leaves, stems and roots.

The ability of desert succulents to hold onto their water content can be attributed largely to their very reduced cuticular evaporation. If cacti are uprooted they will remain turgid (sometimes for many years) without any additional water supply. An uprooted cactus loses about 0.015% 0.05% of its weight each day. If the epidermis is peeled from the plant, nearly 90% of tissue weight is lost in 48 hours. The average ratio of cuticular:stomatal transpiration is 1:5 to 1:50 in desert plants. In plants from moister habitats the ratio is 1:2 to 1:5.

The reduced cuticular transpiration of desert plants is partly due to an increase of surface lipids. The cuticular wax increases at least tenfold under desiccating conditions. It should also be pointed out that as a cuticle dries, the submicroscopic channels passing through it constrict, so that cuticular water loss is much reduced under dry conditions.

The stomatal mechanism is essential for water conservation, and water-saving desert plants spend much of the time with their stomata closed. In addition, water savers can close their stomata very quickly (in as little as 5 minutes in some species). Barrel cacti do not open their stomata at all after 40 days without rain when the water potential of the soil is below that of the plant.

In most plants abscisic acid (ABA), a plant hormone, accumulates in dehydrating leaves. ABA accumulation induces stomatal closure before leaves lose enough water for wilting and passive closure of the guard cell pores. Desert plants exposed to low humidity air close their stomata before dehydration-induced accumulation of ABA occurs. There is some unknown humidity sensing mechanism in these plants.

Although the stomata of desert plants close quickly and stay closed during periods of drought, gas exchange is essential for growth. Therefore stomata must open at some time. To reduce water loss from open stomata the pores of desert plants are frequently located in pits or grooves (Fig. 68). The effect of protecting pores in this way is to interpose between the diffusing stoma and the dry, outer air an air space which is saturated with water vapor. This slows down the rate of water diffusion (and CO_2 exchange). It has been calculated that in *Agave* stomata (see Fig. 68) the superimposition of a pit reduces water loss from the open pore by 31%. In many grasses from arid habitats the leaves fold into rolls that enclose the stomata (Fig. 69). The space inside the fold becomes saturated with water vapor, and this reduces transpiration.

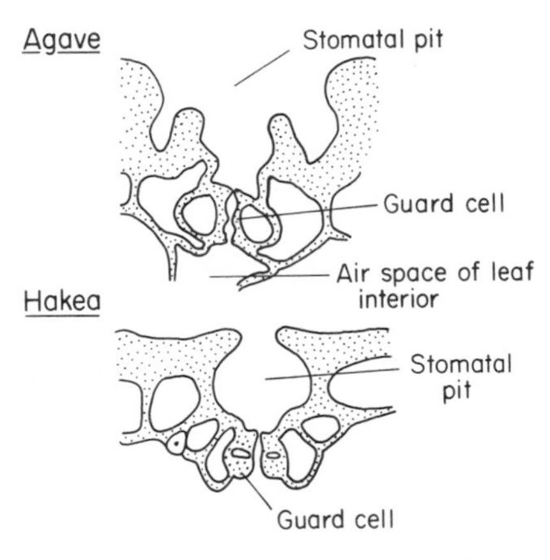

Figure 68. Sunken stomata of two plants from arid regions (modified after Skene, 1924).

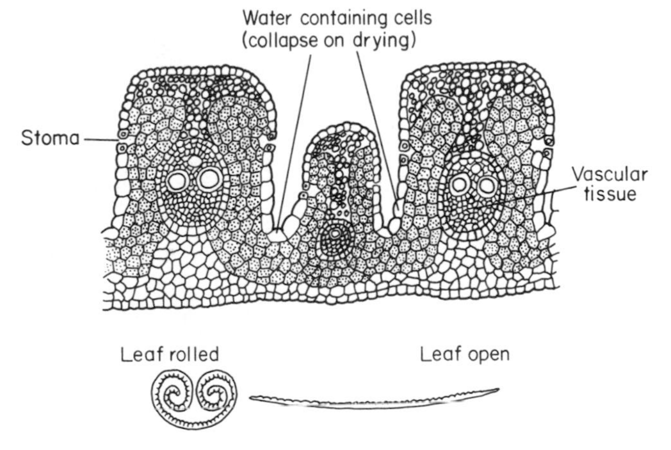

Figure 69. Cross sections of a xeromorphic beach grass leaf that rolls up under desiccating conditions (modified after Parker, 1968).

In some water-saving desert plants transpiration loss per unit surface area is quite high. A strategy of reduced surface area has been adopted in many cases. Extreme examples are seen among succulents that have no photosynthetic leaves. The assimilatory role has been transferred to the stem (which is green). The plants often assume a spherical or barrel-like form that provides a very reduced surface to volume ratio. The leaves of many succulents form spines or thorns (Fig. 70) which may deter grazers. Water loss is very high from tissues exposed to the atmosphere without cuticular protection. Grazing damage could be disastrous in desert conditions.

The commonest method of reducing transpiring surface area is leaf fall. Trees with leaf fall in summer are characteristic of many tropical and subtropical forests. Leaf abscission in autumn is a feature of high latitude forests which must endure a winter drought due to frozen soil. It is interesting that evergreen trees and shrubs of high latitudes have many features associated with reduced transpiration (e.g., thickened cuticle, protected stomata) even in high rainfall areas.

The stems (Fig. 70) and leaves (Fig. 71) of desert succulents serve for water storage. Frequently storage occurs in undifferentiated parenchyma cells. Water storage tissue in the form of enlarged epidermal cells is probably the most frequent adaptation.

The amount of water stored by succulents can be very large. Specimens of the giant cacti of the American southwest contain over 5000 kg of water. The stem surfaces of many cacti are pleated to

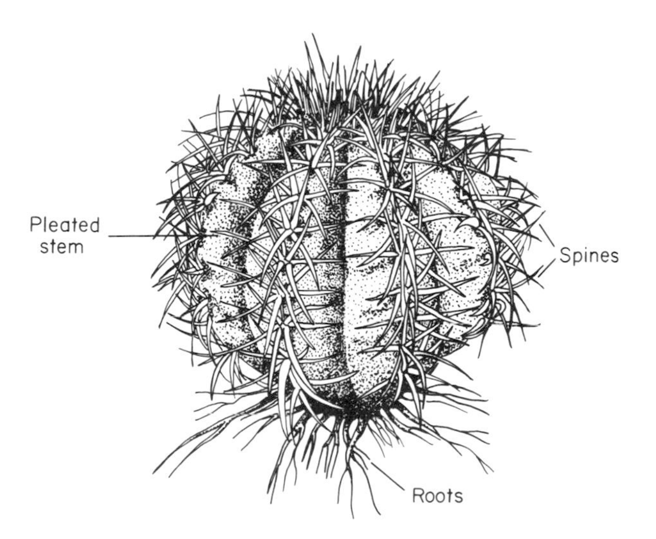

Figure 70. The desert succulent *Echinocactus* showing spine-like leaves, bulbous stem and pleated margins (modified after Goebel, 1889).

allow expansion as water is absorbed and stored (Fig. 70). Whenever water is stored in special tissues they must give up their reserves to assimilating cells. In *Peperomia* the cells of the water tissue collapse while the chlorophyll containing cells (to which water is transferred) remain turgid (Fig. 72). Water storage may occur in root tissues. Huge bulb-like swellings occur in the roots of many desert plants (Fig. 73). During periods of drought in the Kalahari desert (southern Africa) these roots are dug up by the bushmen. The roots are pulped, squeezed and the extracted juices drunk.

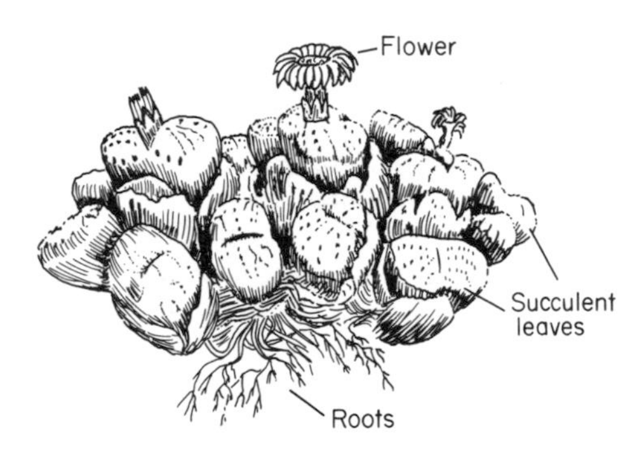

Figure 71. *Mesembryanthemum*, a succulent with water-storing leaves (modified after Goebel, 1889).

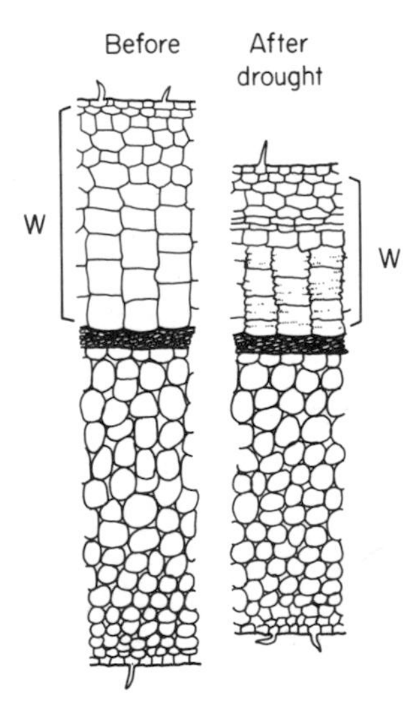

Figure 72. Leaf of *Peperomia* (transverse section) before and after drought, showing shrinkage of water storage tissue while assimilatory cells stay turgid (modified after Haberlandt, 1914).

Figure 73. *Pelagorium* from South Africa showing the bulbous root used for water storage (modified after Gessner, 1956).

Apart from water storage, roots play other important roles in drought avoidance by water-saving desert plants. Water absorption is by roots alone. Nearly all water savers have a root system that is extensive, but very shallow. In *Opuntia*, for example, the anchoring root penetrates to only about 30 cm depth, while the mass of shallow absorbing roots can spread out over 5 m horizontally from the base of the plant. The shallow roots absorb water from desert rains that rarely penetrate the soil to any appreciable depth.

As soils dry up during drought the water potential becomes sufficiently negative that the roots of plants may become major organs of water loss. The plants deal with this problem as they do with water loss from leaves: (a) by making roots impermeable, (b) by abscission and (c) by reducing the root surface area. Obviously the absorbing surfaces of roots cannot be waterproofed, so water savers have developed **rain roots** which develop within a few hours of a rain shower. As the soil dries the absorbing roots are shed. Contrary to expectations, the ratio of root area:aerial organ area may be smaller in extreme water savers than in plants from moist habitats.

Another way in which roots may aid in water retention is through increased hydraulic resistance. It has been suggested that water savers have elevated root resistance to water flow. This would decrease flow to the shoot and reduce water loss from the leaves.

The mechanisms by which water-saving plants reduce the risk of drought have serious assimilatory consequences. Reduction of water vapor loss from aerial organs inevitably reduces CO_2 exchange. This will clearly affect photosynthetic rates. Reduction in surface area: volume ratios will reduce the light-intercepting area of the plant, with obvious consequences. Restriction of root surface area will influence mineral nutrient status. Hence it is not surprising that cacti grow very slowly (only by 10% per year in barrel cactus).

Water Spenders

Because of the relationship between photosynthetic rate and CO_2 exchange rate, the only method of maintaining a high growth rate is for plants to keep their stomata open in daylight. This is what **water spenders** do, and they have much higher growth rates than water savers. Since the plant cannot synthesize water, the absorption rate by the roots must equal the evaporation rate from the aerial organs. Water spenders must have access to large quantities of water. To some extent the supply available to each plant is enhanced by the wide spacing of individuals (so each plant has access to a large soil surface). In addition, desert plants frequently occur in depressions that receive considerable quantities of run-off water. Water spenders have adaptations of their root and conducting systems that permit a massive supply of water to the transpiring surfaces. The most obvious adaptation is the development of a deep root system that extends to the water table. In the Dead Sea area of the near East there are plants with roots that extend to 15 m depth. *Acacia* trees near Suez have roots that penetrate to 30 m. The roots grow down tight fissures in solid limestone, dissolving rock as they go.

Water savers have a high root:aerial organ ratio. A small bush of one desert shrub, *Leptadenia pyrotechnica*, may have a root system penetrating through 850 m^3 of soil. Obviously this gives the relatively small, but rapidly transpiring, aerial organs access to a huge supply of soil water. The massive root systems of water spenders continue to grow during drought conditions that may reduce, or stop, shoot growth. In this way new water supplies can be obtained from unexplored soil.

Water-spending plants have a reduced resistance to water flow in their tissues. Reduced resistance is achieved by increasing the numbers or diameters of xylem tracheary elements. Some xerophytic plants have a greater length of veins per unit area in the leaves than plants from moist conditions.

Absorption of atmospheric moisture (dew, fog or rain) is of minor importance to higher plants. However, for some groups of bryophytes (see Chapter 11), the atmosphere provides a major water supply. When the soil water supply rate is so limited that it is not possible to maintain high transpiration rates, water spenders become water savers. The stomata shut down and the heavy cuticle reduces non-stomatal transpiration.

9

Reproduction in Vascular Plants

The types of reproductive mechanisms employed by extant algae are not at all suitable for life on land. In the sea reproductive spores and gametes are broadcast, in immense numbers, as fragile, unprotected cells. These minute structures have no need of protection in the benign physico-chemical environment of the oceans. On land reproductive disseminules are always in danger of drought once they have been shed from the protective mantle of the parent plant body. A spore can be protected by a waterproof jacket. However, such protection is not available to gametes, which must fuse through naked cell membranes. Vascular plants have evolved fascinating reproductive systems to deal with these problems.

Homospory

There is a basic life history common to all vascular plants. The simple cycle found in ferns serves as an introductory example (Fig. 74). There are two phases in the life history, the **sporophyte** (bears sporangia) and the **gametophyte** (bears gametangia). The gametophyte is diminutive in comparison with the sporophyte (the "dominant" generation).

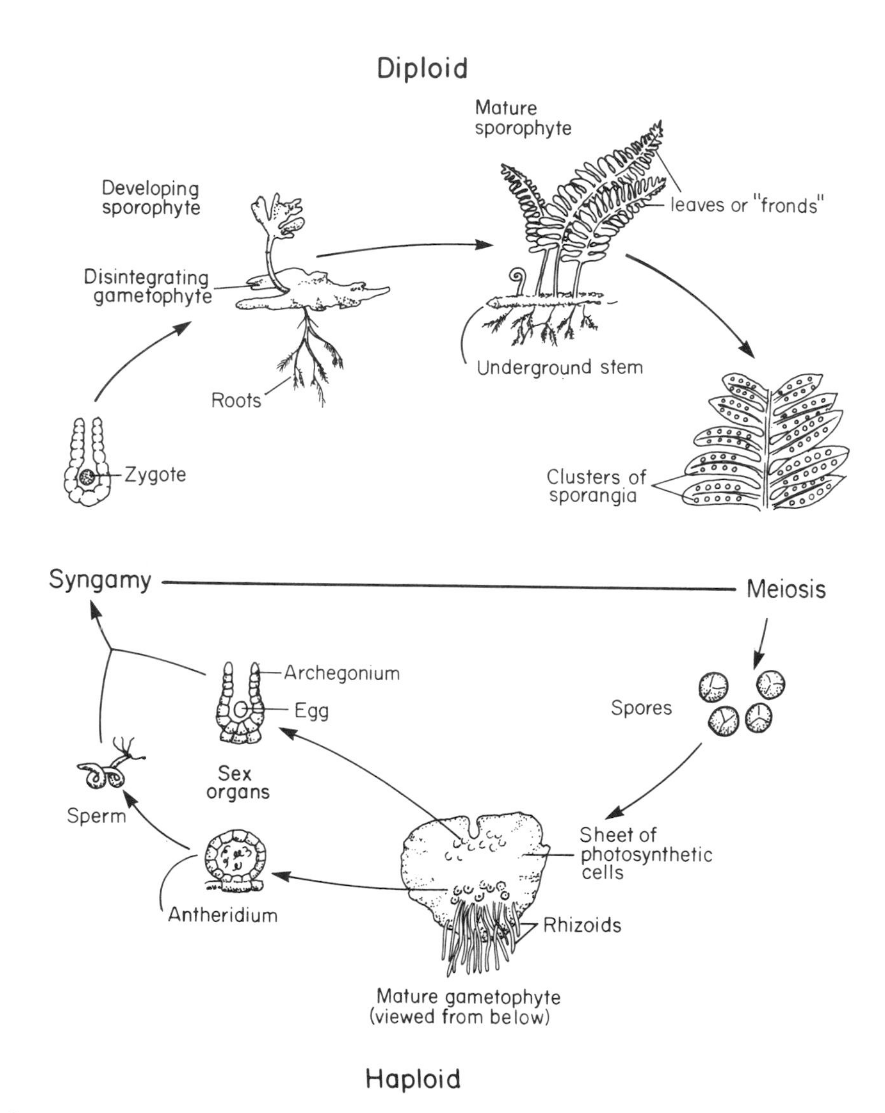

Figure 74. Elements of the life history of a homosporous fern.

In the fern shown in Fig. 74 the sporangia develop on the undersides of leaves. The leaves that bear sporangia in ferns, and all vascular plants are called **sporophylls**. The sporangia of ferns are elaborate, multicellular structures (Fig. 75). Notice that the spores are contained within a sterile jacket of cells. This is a characteristic not found in seaweed sporangia. Apart from the spores and the sporangial wall, most sporangia of vascular plants develop a **tapetum** during early or middle stages of development. A tapetum is a layer of cells inside the sporangium wall surrounding the developing spores. The tapetum is

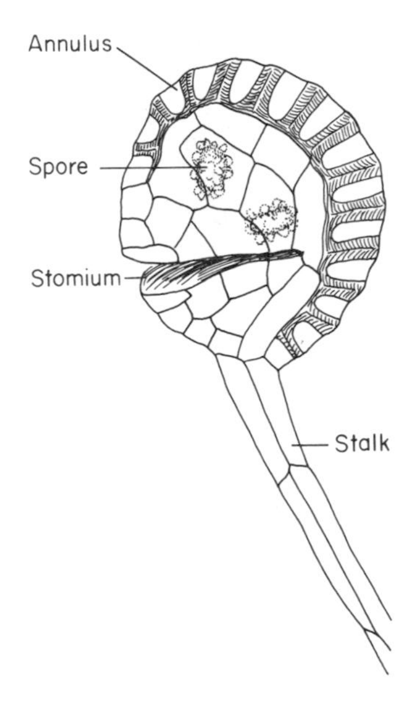

Figure 75. Complex sporangium found in the majority of ferns (modified after Foster and Gifford, 1974).

a nutritive layer that provides food to the spores.

The spores are produced meiotically and are thus haploid. Mature spores of vascular plants have cell walls made up of an extremely complex polymer of cellulose, xylans, lignins and lipids that is called **sporopollenin**. The composition of this substance is similar to that of cutin found in the waterproof cuticle of the vegetative sporophyte. Sporopollenin is probably the most chemically resistant substance produced by plants.

The sporangia of most ferns are specially adapted for dispersal of spores in a dry atmosphere. The sporangium wall contains a chain of cells with thickened cell walls called the **annulus** (Fig. 75). As the sporangium matures, water is lost from the annulus cells by evaporation. There is a powerful adhesion between cell walls and water within the cells. The diminishing water content of the cells causes shrinkage which the walls resist. Because of the uneven thickening of the annulus walls, shrinkage bends the whole band backward so that the sporangium is ripped open (Fig. 76). The annulus is under tremendous tension as the water content of its cells continues to diminish. Eventually the water in each cell breaks away from the cell walls and the annulus springs back into its original position, shooting the spores out into the atmosphere.

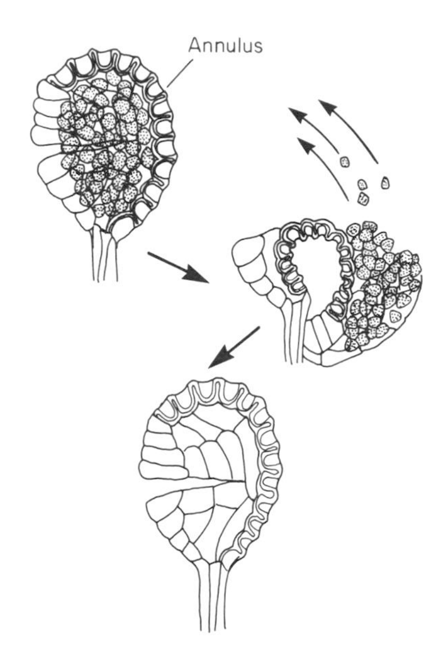

Figure 76. Dispersal of spores from a fern sporangium (modified after Foster and Gifford, *Comparative Morphology of Vascular Plants*. W. H. Freeman & Co. © 1974).

The adaptive features of sporophyte reproduction can be summarized as (a) protection and nourishment of the developing spores by surrounding jackets of sterile cells, (b) the development of a resistant sporopollenin spore wall and (c) explosive aerial dispersal mechanisms.

Spores of all ferns develop into gametophytes provided there is sufficient moisture and light. There is a wide variety of morphological forms among gametophytes. Some are simple filaments that look like green algae. Others are parenchymatous pads of tissue. Whatever the form of the gametophyte, it produces gametangia called **archegonia** (egg producing) and **antheridia** (sperm producing) [Fig. 77]. The gametangia of vascular plants have sterile jackets that protect the developing gametes. All vascular plants are oogamous and the egg cell is never released from the archegonium. In ferns and all the lower vascular plants the unprotected sperms (which are flagellate) must swim to the archegonium in order to effect fertilization. There is an absolute requirement for water in the transfer of gametes, and this must be regarded as a considerable disadvantage for life on land.

After fertilization the egg develops quickly into an embryo enclosed in the archegonium. The young embryo grows out into a new sporophyte (Fig. 74), thus completing the life history cycle.

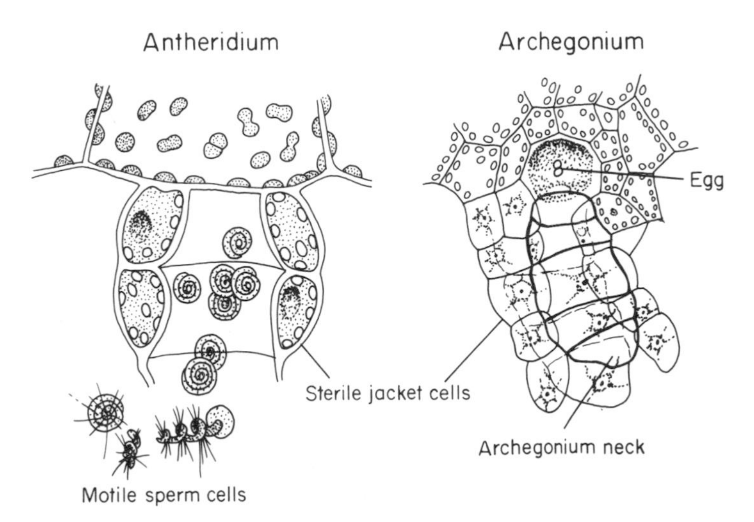

Figure 77. Gametes and gametangia of a fern gametophyte (adapted from *The Plant Kingdom* by W.H. Brown, © 1935, by William H. Brown. Used by permission of the publisher, Ginn and Company [Xerox Corporation]).

The fern life history described above is called **homosporous** because only one type of spore is produced. Each spore develops into a bisexual independent gametophyte.

Heterospory

In a few lower vascular plants and in all the flowering plants and gymnosperms the life history is **heterosporous**. The simplest heterosporous cycles occur among the clubmosses (Lycopsida; see Chapter 10). The sporophyte produces two kinds of spores called **microspores** and **megaspores** (Fig. 78). The names are unfortunate, but have a long tradition of usage. As would be expected, the leaves that bear **microsporangia** are called **microsporophylls**, and those that bear **megasporangia** are called **megasporophylls**.In some clubmosses the sporophylls look like ordinary leaves, except that they bear sporangia. In other types the sporophylls are clustered together on a differentiated axis called a **strobilus** (cone).

The important difference between microspores and megaspores is not their relative size, as the names would suggest. Microspores produce only male gametophytes that produce, in turn, sperms. Megaspores produce only female gametophytes that produce eggs. The male gametophytes of heterosporous forms develop entirely within the spore wall and consist of nothing more than single antheridia.

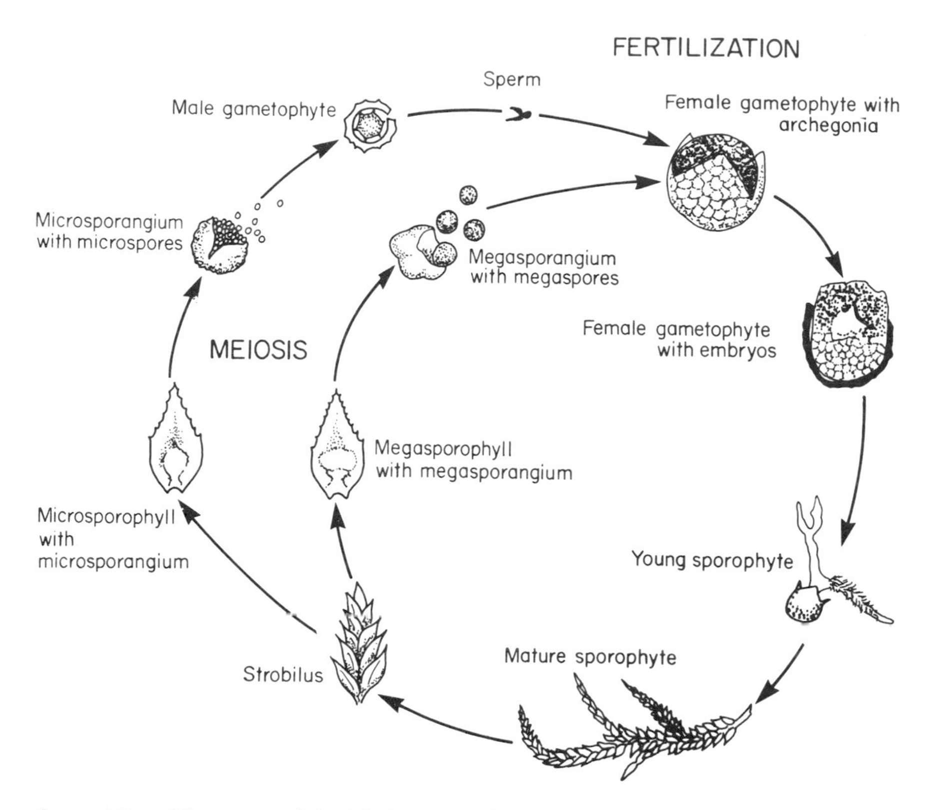

Figure 78. Elements of the life history of an heterosporous clubmoss, *Selaginella* (modified after Greulach, 1973).

The male gametophyte is non-photosynthetic.

In clubmosses the megasporangia shed their megaspores. A female gametophyte develops within the megaspore wall. This wall eventually splits exposing part of the female gametophyte. The exposed part produces archegonia. Fertilization occurs when the microspore wall breaks open, sperms are released and attracted to the egg cell in the archegonium where cellular penetration occurs.

Heterospory is regarded as an essential first step in the evolution of the seed habit. The first vascular plants were certainly homosporous, and all seed plants are heterosporous. The important features of heterospory in club mosses in relation to the development of the seed habit are:

(a) The differentiation of male and female gametophytes,

(b) The development of gametophytes within spore walls and

(c) Nutrient reserves provided by the sporophyte for the development of the gametophytes (the gametophytes do not produce sufficient photosynthate to meet growth requirements).

Seed Habit

All of the flowering plants and gymnosperms (conifers and their allies) produce seeds, and are thus called seed plants. A seed consists of a **seed coat** (diploid, produced by the parent sporophyte) enclosing a female gametophyte (haploid, produced from a megaspore) surrounding a diploid sporophytic embryo (produced by fertilization of the egg cell in the gametophyte; see Fig. 79). The seed can thus be regarded as a structure in which three generations (parent sporophyte, gametophyte and embryo sporophyte) have been telescoped together. The elements of the fern life history have been preserved, but the gametophyte is very reduced and no longer free living.

The seed is important for dispersal and nourishment of the embryo. Of equal significance is the fact that, associated with the production of seed, is an apparatus that frees vascular plants from dependence on water for reproduction. The life history and reproduction of pine will illustrate the salient features of reproductive adaptions in seed plants (Fig. 80). Two types of cones are produced by pine trees. The seed cones are made up of scales (called **ovuliferous scales**) each of which bears two megasporangia. Each megasporangium is enclosed within an **integument**—an enclosing structure developed by the parent sporophyte (Fig. 80). In seed plants the

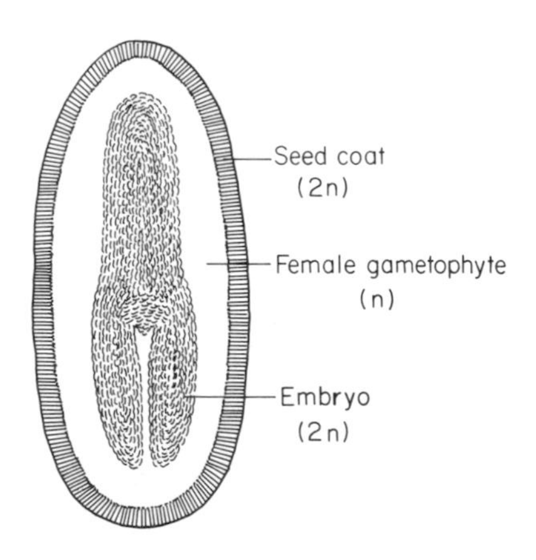

Figure 79. Components of a gymnosperm seed (modified after Foster and Gifford, *Comparative Morphology of Vascular Plants*. W. H. Freeman & Co. © 1974).

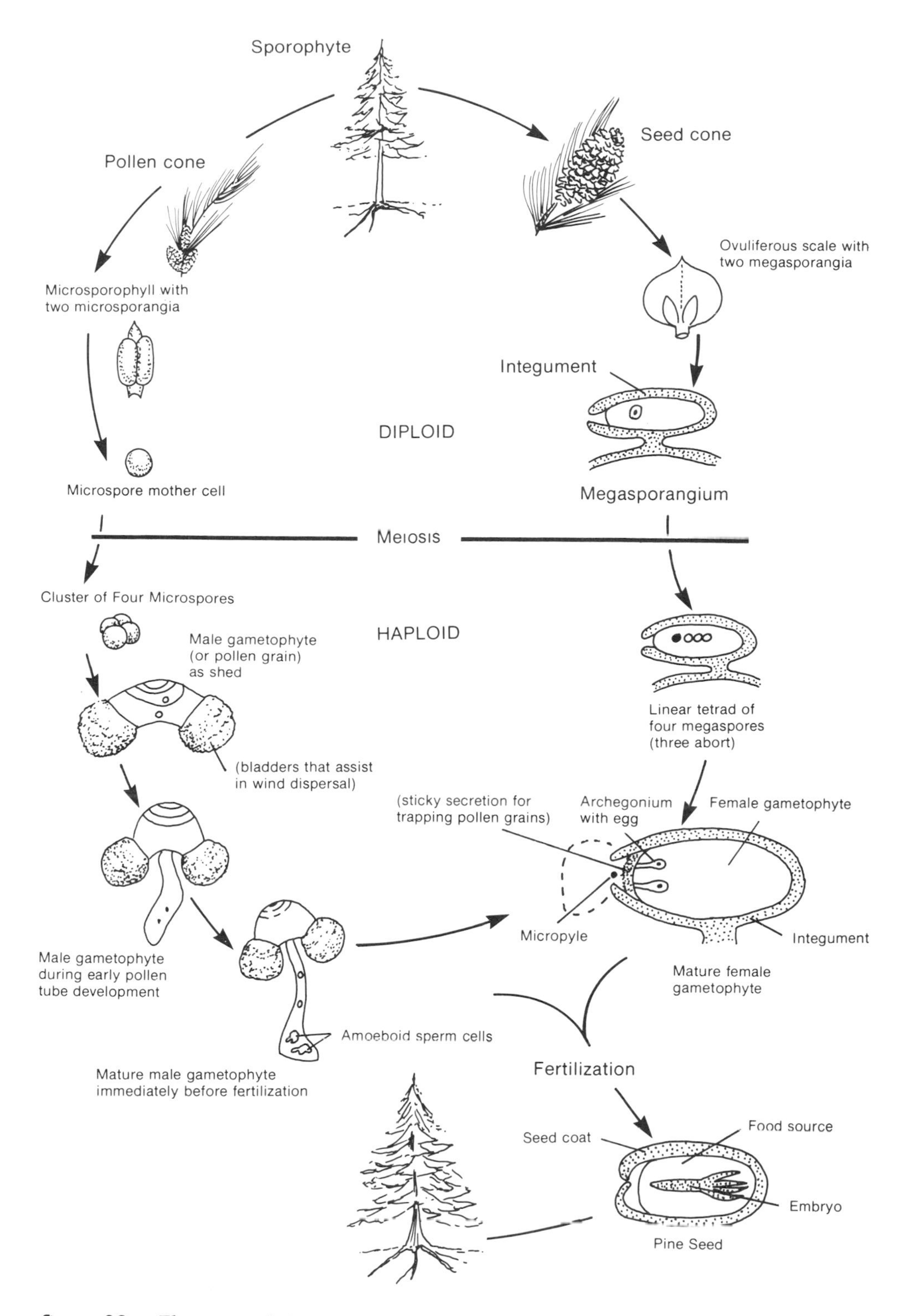

Figure 80. Elements of the life history of a pine tree.

megasporangium and its integuments are called the **ovule**. To further complicate the terminology, the megasporangium of seed plants is called a **nucellus**.

Within the megasporangium a functional haploid megaspore is produced meiotically. This spore develops into a female gametophyte that is retained within the sporangium wall. The female gametophyte is never released from the scales of the seed cone. Archegonia develop at the end of the female gametophyte facing the **micropyle** which is a small tube in the integument allowing access to the outside world.

The pollen cone has microsporophylls that bear microsporangia (Fig. 80). Large numbers of microspores are produced in each sporangium. Before the microspores are shed, a male gametophyte develops within each spore wall. The binucleate male gametophyte is enclosed within a spore wall that develops two large bladders to assist in wind dispersal. The entire structure is called a **pollen grain** and is not fundamentally different from the endosporic gametophyte of some clubmosses.

If a pollen grain lands on the micropylar end of an ovule, the grain germinates. A narrow tube grows down through the micropyle. This is the **pollen tube** (unique to seed plants). Two sperm cells pass down the pollen tube and are injected into an archegonium. Fertilization results in the development of an embryo that is nourished by food stored in the female gametophyte. The composite structure is, of course, a seed.

The life history of a conifer is a simple modification of that which is found in ferns. There are sporophytes, gametophytes, sporangia, archegonia, and so on. Modification has occurred through the extreme reduction of the gametophytes. The male gametophyte is so reduced that no gametangia are formed and motile sperms are dispensed with. The male gamete is entirely protected by the microspore wall. Final transfer to the egg occurs through the pollen tube, a structure not found in lower vascular plants. The second unique structure of seed plants is the integument which encloses the megasporangium, female gametophyte and, ultimately, the developing embryo.

The reproductive mechanisms of seed plants are highly adaptive to desiccating terrestrial conditions. Fertilization is internal so that gametes are never exposed to drought. Internal fertilization is made possible by the extreme reduction of the male gametophyte so that it exists only within the protective jacket of the microspore wall. The development of pollen grains was made possible through the evolution of heterospory from the primitive homosporous condition.

In seed plants the egg is protected within the megasporangium.

Again this is made possible by the extreme reduction of the gametophyte (female) and the occurrence of heterospory. Following fertilization the embryo is nourished by the parent sporophyte and gametophyte and is dispersed in a compound structure called a seed.

Flowering Plants

Flowering plants are, of course, flower bearing. The flower and other components of reproduction seem quite exotic. However, close examination reveals a life cycle that is clearly related to those of all other vascular plants.

The parts of a flower are shown in Fig. 81. The sporangium bearing structures are called **stamens** and **carpels**. This terminology was introduced before the homologies with other vascular plants were fully appreciated. The tradition of usage has continued and is followed here. The stamens produce microspores in terminal, swollen **anthers**. The ovules (megasporangia+integuments) are enclosed within the swollen bases of the carpels called **ovaries**. The ovary bears a **style** with a terminal **stigma** that receives pollen grains. **Sepals**

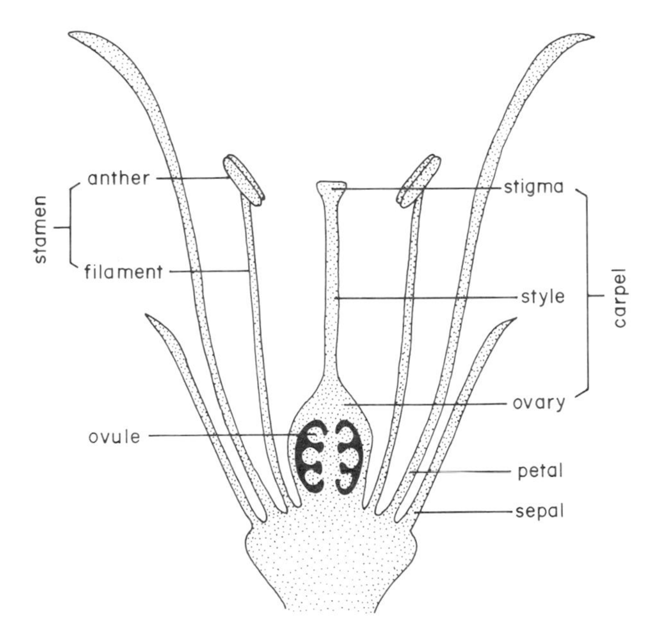

Figure 81. Parts of an angiosperm flower in longitudinal section (modified after Delevoryas, 1966).

and **petals** are often showy appendages that attract animal pollinators (pollen vectors). Sepals and petals are derived leaves.

There is reason to believe that the stamens and carpels of flowering plants are, in fact, highly modified sporophylls. Among modern forms there is a species, *Drimys piperata*, that has a carpel which is clearly a folded leaf-like structure (Fig. 82). Leaf-like stamens can also be found (Fig. 83). Although there is considerable variation in the details of the reproductive cycle in angiosperms, a typical plant would go through the stages shown in Fig. 84. Each ovule within the ovary of a carpel consists of a megasporangium enclosed within an integumentary system (two-layered) supported by a stalk. Following

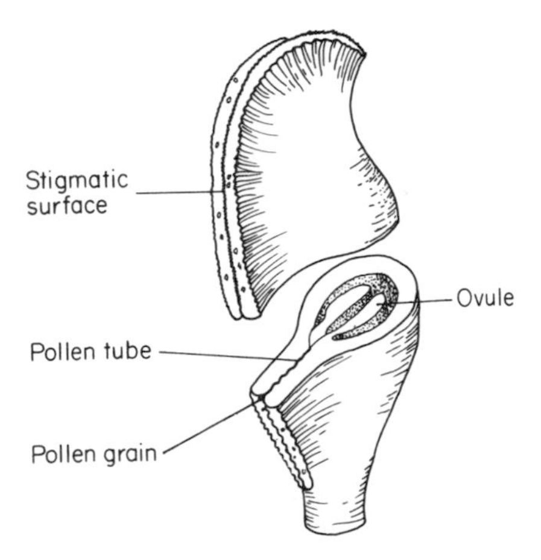

Figure 82. Carpel of *Drimys piperata* showing folded leaf-like construction (modified after Savidge, 1976).

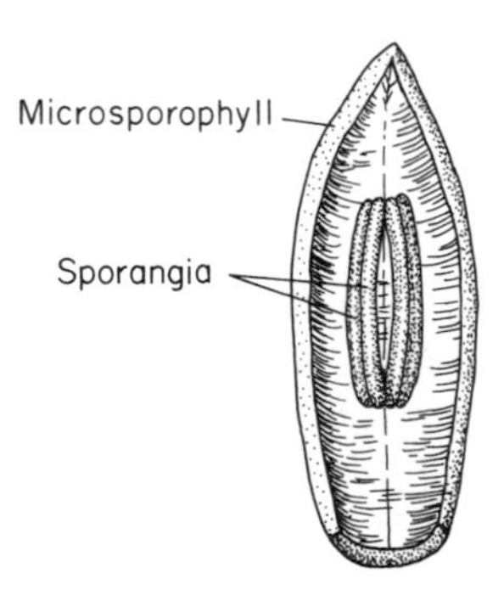

Figure 83. Leaf-like stamen of *Austrobaileya maculata* (modified after Canright, 1952).

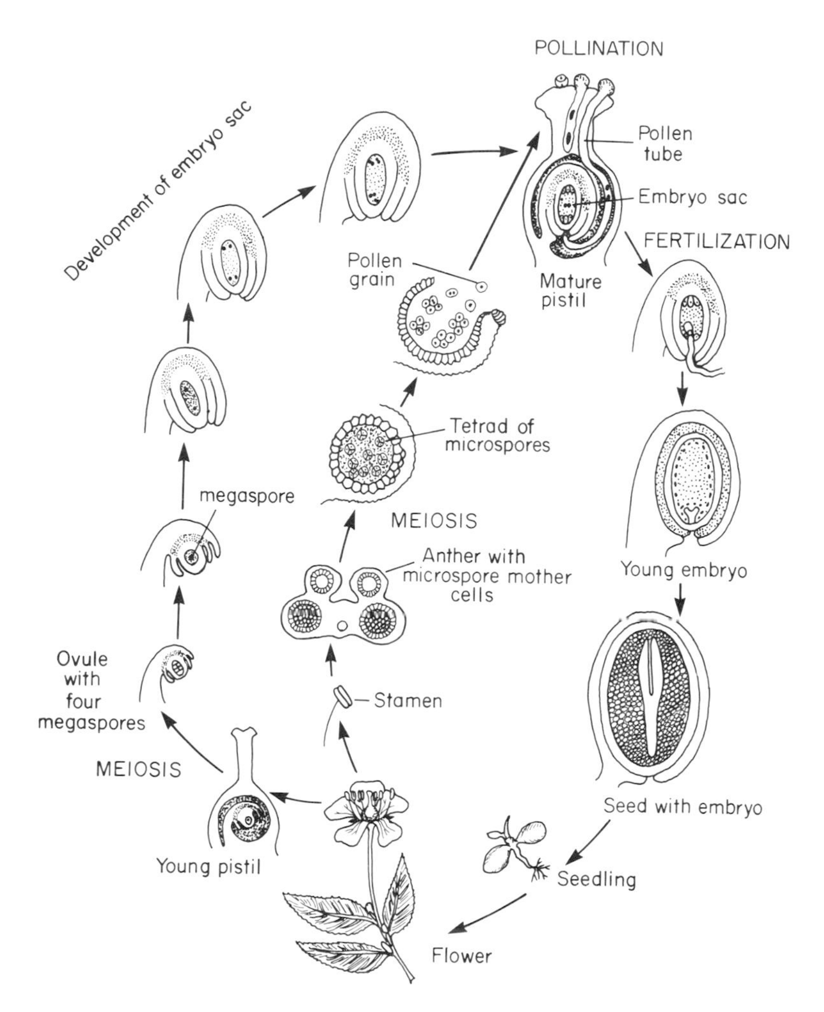

Figure 84. Elements of the life history of a typical angiosperm (modified after Sinnott, and Wilson, *Botany, Principles and Problems.* © 1963 McGraw-Hill Inc., New York).

meiosis one viable megaspore develops into the female gametophyte which is called an **embryo sac** in angiosperms.

The female gametophyte commonly consists of just eight nuclei. Two of the nuclei fuse to form a diploid **endosperm nucleus**. One of the remaining haploid nuclei acts as an **egg nucleus**. Archegonia do not develop in angiosperms. In fact, gametangia are entirely absent in the group.

The formation of the endosperm nucleus is a very distinctive feature of angiosperm female gametophytes. It is not found elswhere in the plant world.

The stamens bear microsporangia that produce microspores. Before release, the spore nucleus divides into two cells called a **tube cell** and a **generative cell** (Fig. 85). The male gametophyte inside the spore wall is the pollen grain.

By various vectors (discussed below) pollen is transfered to the stigma of the carpel. Germination of the grain occurs and a long pollen tube pushes down through the tissues of the carpel. As the tube grows the generative nucleus divides into two sperm cells. At maturity the male gametophyte consists of three cells.

The pollen tube grows through the gaps in the integument of one of the ovules (Fig. 84). One of the sperm cells fuses with the egg cell to form the diploid zygote nucleus. The other sperm cell fuses with the endosperm nucleus to form a triploid nucleus. This gives rise to a triploid endosperm tissue in the seed that surrounds and nourishes the embryo.

The basic features of all vascular plant reproduction are to be found in angiosperms. The double integumentary system of the ovule and the triploid endosperm tissue are structures that are unique to flowering plants.

FRUITS

Many angiosperms develop fruits that are used for seed dispersal. The vectors may be animals, or use may be made of abiotic agents such as wind and rain. A fruit develops by enlargement and modification of the carpels and may include accessory parts of the flower. In the apple fruit (Fig. 86) the core is derived from the flower ovary and contains the seeds. The fleshy part of the fruit is derived from the fused bases of the stamens, sepals and petals.

There is an enormous variety in the structure and dispersal mechanisms of fruits. They are all, however, derived from carpels and flower accessory structures. This unique development is confined to angiosperms.

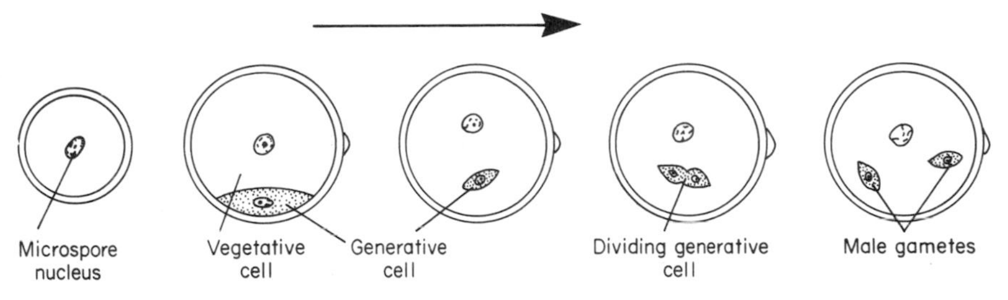

Figure 85. Development of an angiosperm pollen grain (modified after Savidge, 1976).

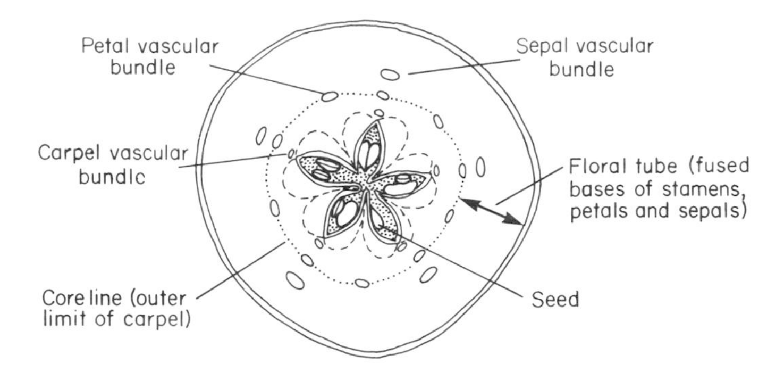

Figure 86. Parts of an apple fruit in cross section (modified after Robbins et al., 1964).

FLORAL MECHANISMS

Although a flower is made up of many parts, the parts are designed to act together to ensure that the following processes are carried out:

(a) Cross-pollination and

(b) Interspecific reproductive isolation.

Floral mechanisms designed for cross-pollination prevent certain gametes of the same species from combining. Reproductive isolation exists when crossing is prevented between members of different species. Floral mechanisms that prevent interspecific crosses represent only one of at least ten different isolation barriers that may occur in angiosperms. Included among these barriers are geographical isolation, hybrid sterility and so on.

The discussion that follows deals only with floral devices in relation to cross-pollination and interspecific isolation. It must be said, first or all, that the subject is very complicated, but it is very interesting and has fascinated generations of naturalists. To begin with, cross-pollination is *not* synonymous with cross-fertilization. Cross-fertilization occurs when two different gametic genotypes fuse to produce a viable zygote. Cross-pollination occurs when pollen is transferred between two different flowers. The flowers may be on the same individual organism or on different organisms. Only pollination between different individuals produces cross-fertilization. Pollination between flowers on the same individual is genetically equivalent to self-fertilization.

The importance of pollen transfer between two individuals is that two genotypes combine on gamete fusion. This leads to the production of genetic variation. The rate of evolution (change in population gene frequencies) is proportional to the amount of genetic variation in a population. By any standards the rate of evolution among angiosperms is very high. The rate of evolution of these plants is likely related to the precision of pollen transfer.

The floral devices that promote pollen transfer between two individuals can be classified as follows:

(a) All flowers identical; presentation and reception of pollen in the individual flower separated in time,

(b) All flowers identical; stamens and carpels spatially separated in each flower,

(c) Stamens and carpels in separate flowers on the same plant or on separate plants,

(d) Two or more different flower morphologies, but all flowers with carpels and stamens.

All of these mechanisms may occur in combination and in association with other barriers to self-fertilization.

In case (a) [all flowers identical] the carpels and stamens mature at different times. This ensures that self-pollination cannot occur. Depending on the species, stamens or carpels mature first. Generally, earlier maturation of carpels is a more effective barrier to self-pollination. Once a carpel has received sufficient pollen to ensure fertilization of all egg nuclei, self-pollination from later maturing stamens in the same flower is impossible. However, when stamens mature before carpels, all the pollen must be removed from the flower before the carpels become receptive.

Case (b) [identical flowers, spatial separation of stamens and carpels] flowers are the most familiar and common forms among angiosperms. The stamens and carpels mature together within the same flower. However, each flower is structured so that the pollination vector will not transfer pollen between stamens and carpels of the same flower. The elaborate devices employed by orchids, for example, to effect cross-pollination are highly ingenious and fascinating to naturalists.

In case (c) [stamens and carpels in separate plants] the floral devices ensure that self-fertilization does not occur. The separation of staminate ("male") and carpellate ("female") plants is functionally equivalent to the sexual differentiation that is so common in animals. This differentiation prevents half of a population from producing seed. The separation of stamens and carpels in flowers on the same plant has the advantage that all individuals are potentially seed bearing, but this floral mechanism does not prevent self-fertilization.

The functional significance of case (d) floral structures [two or more morphologies, all with stamens and carpels] is unknown. In *Primula* species (primroses) there are two types of flowers called **pin** and **thrum**. The pin flowers have long styles and the thrums have short styles. The flowers differ in anther position. Long-styled flowers have low anthers. Thrum flowers have the reverse combination. There are internal incompatibility mechanisms which ensure a high seed set when pollination occurs between pin and thrum flowers. When a pin flower is pollinated by a pin, or a thrum by a thrum, seed set is reduced. Compatability is largely related to pollen tube development. On the "wrong" stigma pollen tube development is retarded. Whatever the details of incompatability within pin and thrum flowers, the different floral structures do not seem to ensure cross-pollination. The functional role of heteromorphism is not understood.

When the floral mechanisms (cases a-d) occur in combination, there is more or less complete prevention of self-fertilization. The diversity of angiosperm approaches in reaching this objective contrasts startlingly with the uniformity found in other groups of green plants (and animals).

POLLEN VECTORS

Pollen is transferred by wind, water and by animals. The greatest diversity of floral mechanisms occurs in animal pollinated angiosperms. Because insects are the major animal vectors, and because insects are so diverse (most species of living organisms are insects), it is hardly surprising that there is a great diversity of animal pollinated floral devices. It is not possible here to attempt a review of the range of floral strategies employed to ensure cross-pollination by insects. Reviews of this kind normally appear as catalogs of specific case histories without underlying thematic principles. The approach here will be limited to a presentation of the floral **syndromes** associated with wind, animal and water pollination.

Wind pollinated flowers are usually unisexual and mature before leaves come out, or the flowers are exposed outside the leaf mass. Obviously a leaf canopy will interfere with patterns of air movement that disperse the pollen. The petals and sepals of the flowers are small or absent. Attractive colors and scents are absent.The stamens are exposed from the flower so that the pollen can be picked up by wind currents. The stigmas are also exposed so that they can receive wind blown pollen. The pollen grains are small (20-30μm diameter), smooth and dry for easy transport. Pollen is produced in large quantities and is trapped by stigmas that have expanded surfaces. Stigmas appear feathery or brush-like in contrast to the knob-like stigmas of animal pollinated flowers. Although pollen grains may be carried hundreds of kilometers by the wind, the area of massive pollen deposition is quite small (Fig. 87).

Animal vectors visit flowers for food (nectar, pollen, wax or flower parts) or they are deceived by a flower that looks like some other attractive object. An extreme example in this latter category occurs when flowers are modified to appear as mates for visiting insects. Some insects will attempt copulation with the disguised flower. In order to ensure animal visitation the flowers advertise their presence with brightly colored parts and with air carried odors. It is interesting that the sights and smells that attract insects are, in many cases, thought attractive by human observers.

Pollen production, in relation to ovule production, is much reduced in animal pollinated flowers. Pollen grains are often sticky and sculptured so that they stick to their animal vectors. Apart from insects, the commonest animal vectors are vertebrates. The major agents are birds and bats.

There is no general syndrome of water pollination mechanisms. Only a few genera use water currents as vectors. In most cases pollination occurs on the water surface. Pollen (or pollen bearing flowers) are released on the water surface and carried to receptive flowers.

The precision of pollen transfer is greatest when animal vectors

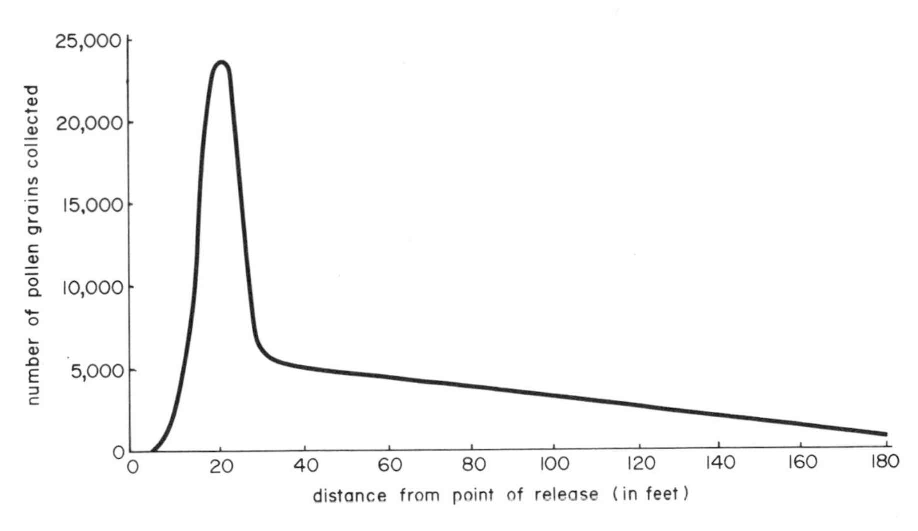

Figure 87. Numbers of pine pollen grains collected at various distances from a release point (modified after Grant, 1963. © Columbia University Press. By permission).

are used. It is important to note that wind-pollinated plants frequently occur gregariously, sometimes in monospecific stands. The prairie grasslands and coniferous forest are obvious examples. This gregariousness is essential when the precision of pollen transfer is not great. In contrast, within tropical rain forests there is a huge diversity of widely separated insect-pollinated plants. Because the flowers advertise themselves so well, animal vectors can readily locate widely dispersed plant individuals.

REPRODUCTIVE ISOLATION

There are many plant species that grow together, bloom together, are more or less interfertile, but do not normally produce hybrids in the wild. Interfertility in these cases can be demonstrated by artificial pollination. The reason that hybrids do not normally occur can be found in the floral mechanisms. An example can be found in the columbines of western North America. Two species in this group are shown in Fig. 88. The flower with the short spurs of 1-2 cm length, *Aquilegia formosa*, is pollinated by humming birds with short beaks of 1-2 cm length. The birds obtain nectar from the spur bases. In *A. pubescens* the spurs are 3-4 cm long and a hummingbird could not reach the nectar. The birds do not even attempt to obtain nectar from this species. The long-spurred flowers are pollinated by moths with probosci 3-4 cm long. The pollinators are faithful to each flower species and reproductive isolation is effected.

The importance of reproductive isolation is that it enables species that live together to remain distinct. Hybridization will lead to a single species. Members of different species have different resource

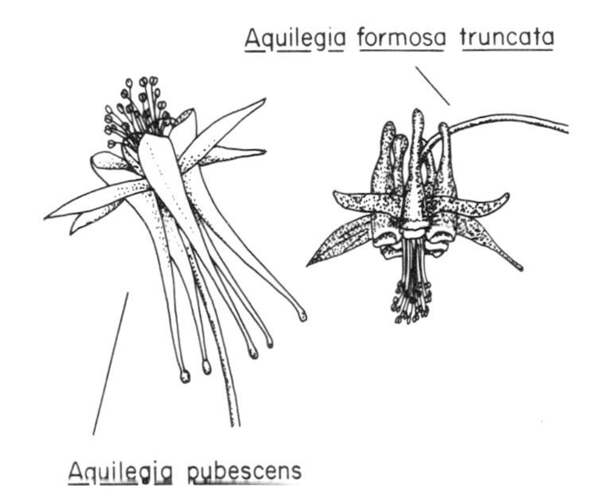

Figure 88. Flowers of two columbine species (modified after Grant, 1963).

requirements. Different species can allocate resources among themselves. Hybridization leads to uniform resource requirements and subsequent resource depletion. There is, therefore, strong selective pressure for the divergence of species. Reproductive isolation allows this to happen.

FLORAL EVOLUTION

It seems likely that flower types co-evolved with their insect pollinators. The dependence of most flowering plants on their insect pollinators makes this probable.

The beetles (Coleoptera) appeared about 400 million years ago and thus preceded the angiosperms. As the angiosperms were diversifying in the Cretaceous the two-winged flies (Diptera) evolved. The four-winged wasps and bees (Hymenoptera) appeared a little later. Highly specialized moths and butterflies (Lepidoptera) evolved about 40-50 million years ago. The flowers of angiosperms are thought to have evolved in concert with the appearance of the major insect pollinator groups. Using modern plants that apparently form a series between primitive and advanced conditions, it is possible to construct a theory of floral evolution (Fig. 89). According to this theory there are six levels of evolutionary progression that can be recognized. The

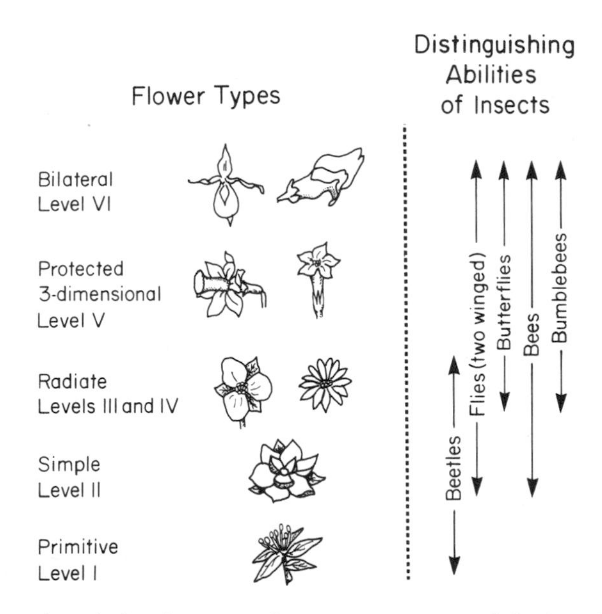

Figure 89. Relationship between flower structure and distinguishing abilities of five groups of insects (modified after Leppik, 1957).

primitive level (I) occurs in flowers that have no special symmetry or colors. Each of the flower parts is duplicated many times. Flowers of this type are pollinated by insects that have no special sensory abilities and use the flower parts as food. Beetles are obvious candidate pollinators. The **simple level** (II) comprises dish like flowers without definite symmetry and with simple colors like yellow, white or green. Among modern plants *Magnolia* and *Liriodendron* (tulip tree) are good examples. Flowers at the simple level are pollinated by beetles and two-winged flies that are foraging for food. Flowers at levels III and IV are called **radiate** types. These have radial symmetry and flowers of single colors (white, yellow, blue, red, etc.). Some members of this group have fixed numbers of floral parts. Insects that pollinate these flowers must be able to recognize two-dimensional radial symmetry and several colors other than white, green and yellow. Flowers at level V are of the **protected** type. These have a three-dimensional tubular structure, nectar glands and a general reduction and fusion of floral parts. The flowers are generally radially symmetrical and have colors similar to level IV flowers. Insects pollinating level V plants must be able to distinguish three dimensional patterns (and this excludes beetles as potential pollinators). Flowers at level VI, the **bilaterally symmetrical** types, all have variegated colors and the finest scents. Orchids, irises and snapdragons fall into this category. Pollinators are restricted to butterflies, moths, bees and birds that can recognize bilaterally symmetrical patterns.

10

Systematic Survey and Evolutionary History of Vascular Plants

Vascular plants belong to a single division, the **Tracheophyta**. The tracheophytes are plants that have xylem (tracheids and/or vessels). They also have a life history in which the gametophyte is diminutive and the sporophyte dominant. The last characteristic distinguishes tracheophytes from bryophytes (mosses, hornworts and liverworts; see Chapter 11), but not from the various seaweed species that have dominant diploid phases. Tracheary elements of xylem seem quite diagnostic for vascular plants, although there have been suggestions that tracheid-like elements occur in the sporophytes of certain bryophytes.

The Tracheophyta has many cytological and biochemical features in common with the "bryophytan" line of the Chlorophyta (Chapter 4). The two groups have the same major accessory photosynthetic pigment (chlorophyll *b*), the same storage product (starch), similar cell wall composition (mostly cellulose) and similar cell division mechanisms. The prevailing view is that the "bryophytan" line of the Chlorophyta gave rise to the vascular plants.

The fossil record provides no clues as to the structure of green algal ancestors of land plants. Green algae appear in the fossil record from about one billion years ago, but for most of eukaryote history members of the Chlorophyta were unicellular or colonial in morphology. Complex filamentous and parenchymatous forms do not appear in the geological record until after the first vascular plants. Among leading authorities there is now a view that vascular plants (and bryophytes) originated from unicellular or colonial green algae that invaded the land.

The major forms of life on land (vascular plants, insects and tetrapod vertebrates) had their origins at approximately the same time—around 400 million years ago. Before that time it seems likely that damaging levels of UV radiation prevented terrestrial colonization. The accumulation of photosynthetic oxygen and the formation of the ozone ultra-violet screen prepared the way for life on land. The landscape of >400 million years ago must have been a barren scene. Yet within 60 million years tree-like organisms reached to heights of 30 m or more. A lush vegetation soon developed and the coal age began. In geological terms, all of this took place "overnight."

The division Tracheophyta is subdivided into eleven classes of which four are extinct (Table 5). Of the seven classes that are extant, four of them (Psilopsida, Lycopsida, Sphenopsida and Gnetopsida) have only eleven genera between them. Most modern vascular plants are either ferns (Filicopsida), gymnosperms (Gymnospermopsida) or flowering plants (Angiospermopsida). The vast majority of plant species in the world belong to the Angiospermopsida, a group that originated only 130 million years ago.

Table 5. Classification of the division Tracheophyta

Class	*Common Name*
Rhyniopsida*	Primitive vascular plants
Psilopsida	Whisk ferns
Zosterophyllopsida*	None
Lycopsida	Clubmosses
Trimerophytopsida*	None
Sphenopsida	Horse tails
Filicopsida	True ferns
Progymnospermopsida*	Ancestors of gymnosperms
Gymnospermopsida	Naked seed plants
Gnetopsida	Gnetophytes
Angiospermopsida	Flowering plants

* Extinct groups

Class Rhyniopsida

Members of this class were among the earliest of land plants, and they are all extinct. They were morphologically very simple, consisting of upright axes with terminal sporangia and horizontal underground stems furnished with absorbing rhizoids (Fig. 90). Leaves and roots were absent. The plants were also anatomically simple. In the simplest, *Cooksonia*, guard cells were absent from the stomata and the plants may have lacked intercellular air spaces. *Rhynia*, another very ancient genus, had fully functional stomata. All the Rhyniopsida had tracheids surrounded by sieve cells, and a cuticle on the epidermis.

All the morphological and anatomical characteristics described

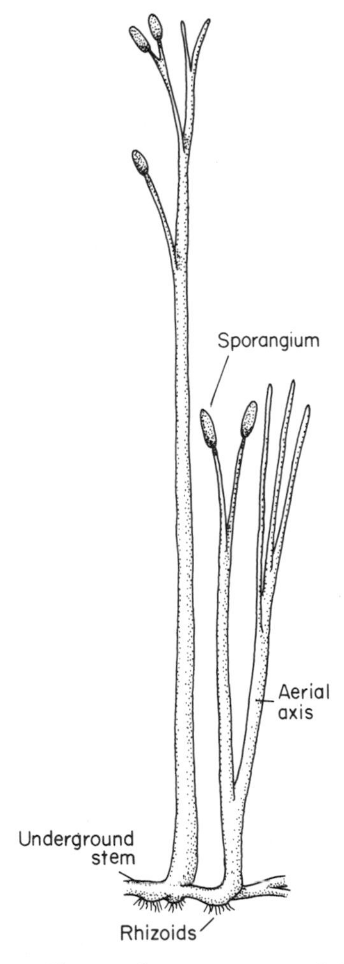

Figure 90. Vegetative and reproductive organs of the sporophyte of *Rhynia* (modified after Kidston and Lang, 1921).

above for this class refer to the obvious sporophyte generation. The nature of the gametophytes is a somewhat contentious issue. Three fossil gametophytes are known from about 380 million years ago. A reconstruction of one of these (*Sciadophyton*)is shown in Fig. 91. Axes of up to 10 cm in length radiate from a flat central structure. The gametangia are held aloft in special funnel-shaped axes. The gametangial axes contain genuine tracheids. Archegonia and antheridia were borne on the gametangial axes. The structure of these early gametophytes is very reminiscent of modern-day bryophytes.

Class Psilopsida

There are two modern genera in this class, *Psilotum* and *Tmesipteris* *Psilotum* has a very simple morphology that is similar to that of *Rhynia*. The sporophyte consists of upright, branched stems that lack leaves with vascular tissue. Small scales are present on the stem. There is an underground stem, but no roots occur in members of this class. Sporangia are borne laterally on small axes. *Tmesipteris* is similar to *Psilotum*, but has vasularized leaves.

The sporophytes of *Psilotum* and *Tmesipteris* are large (20-100 cm tall) in comparison with the gametophytes (several mm in length). Tracheids occur in the gametophytes of *Psilotum*.

Multicellular antheridia and archegonia are scattered over the surfaces of psilopsid gametophytes. Antheridia produce multiflagellate sperms that must swim to effect fertilization.

The morphological and antomical features of the Psilopsida are very similar to those of the Rhyniopsida. The evolutionary relationships of the two classes are unclear because the Rhyniopsida disappeared from the fossil record about 360 million years ago.

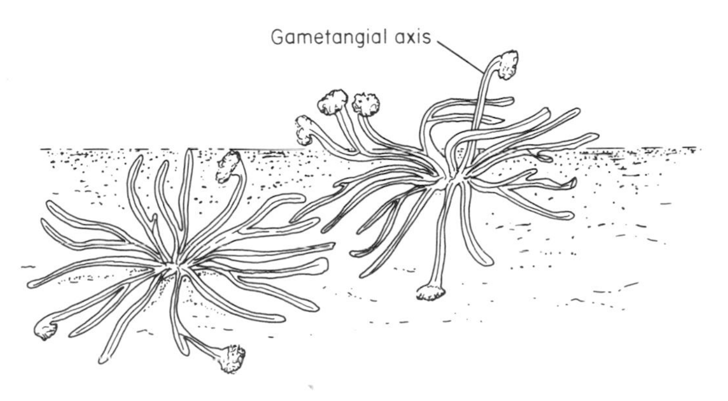

Figure 91. Gametophyte of *Sciadophyton* (modified after Mägdefrau, 1953).

Class Zosterophyllopsida

This class comprises several extinct genera that first appeared in the fossil record from about 395 million years ago. Zosterophyllum (Fig. 92) was a tufted plant that grew to about 15 cm in height. Roots and leaves were absent. Although cuticle and stomata occurred on upper axes, they were absent from the lower portions of each plant, suggesting that the genus was amphibious. The sporangia were borne laterally on terminal spikes. The lateral position of the sporangia, their structure, and the occurrence of a massive xylem sets the zosterophylls apart from the Rhyniopsida in most modern classifications.

Class Trimerophytopsida

Members of this class are extinct and appear in the fossil record from a little less that 395 million years ago. Plants in this class differ from the Rhyniopsida and Zosterophyllopsida in having a definite main axis with more or less determinate lateral branches (Fig. 93). Sporangia were borne on the ultimate branches of the laterals. Leaves and roots were absent in the trimerophytes.

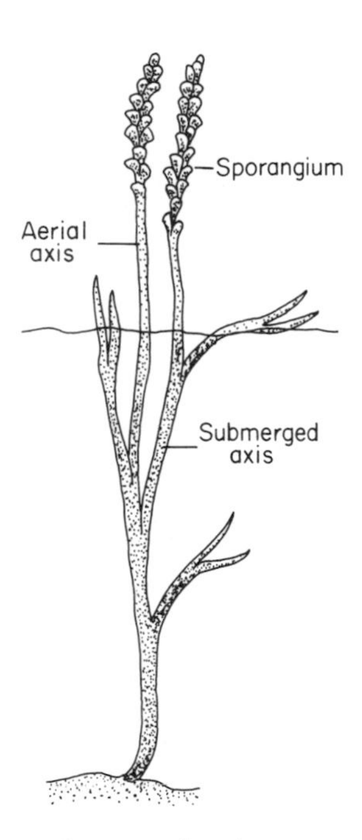

Figure 92. Vegetative and reproductive organs of the sporophyte of *Zosterophyllum* (modified after Stewart, 1983).

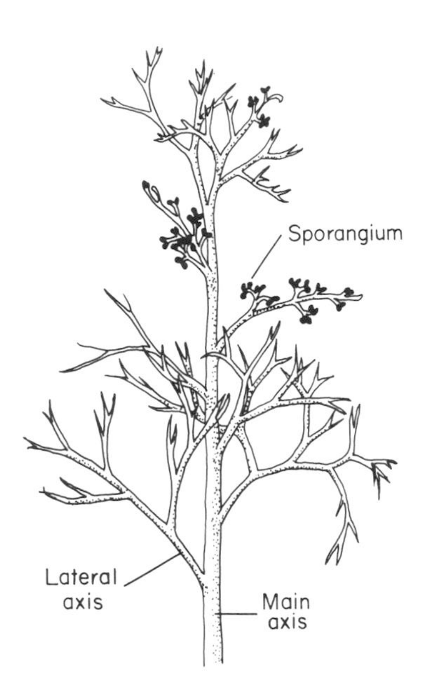

Figure 93. Aerial axes and sporangia of *Psilophyton* (modified after Andrews et al, 1968).

The classes Rhyniopsida, Zosterophyllopsida and Trimerophytopsida are thought to be extinct ancestors of all modern vascular plant groups. There is a prevailing opinion which places the trimerophytes (descendants of the first rhyniophytes) as ancestors of the Sphenopsida, Filicopsida and all seed plants. The zosterophyllytes, also presumed descendants of the rhyniophytes, are thought to be ancestors of the Lycopsida. The evidence for this view is rather tenuous, but it is clear that the ancestors of all vascular plants are to be found among the three classes that colonized the land between 395 and 400 million years ago.

Class Lycopsida

There are five extant genera of clubmosses. The group is very ancient and most of its members are now extinct. The class is distinguished (from the four considered above) by the presence of true roots and leaves (Fig. 94). The sporangia are attached to leaves (called sporophylls).

The first land plants had no roots. Absorption of water and minerals occurred through underground stems with absorbing rhizoids (multicellular hair-like structures). There are major differences

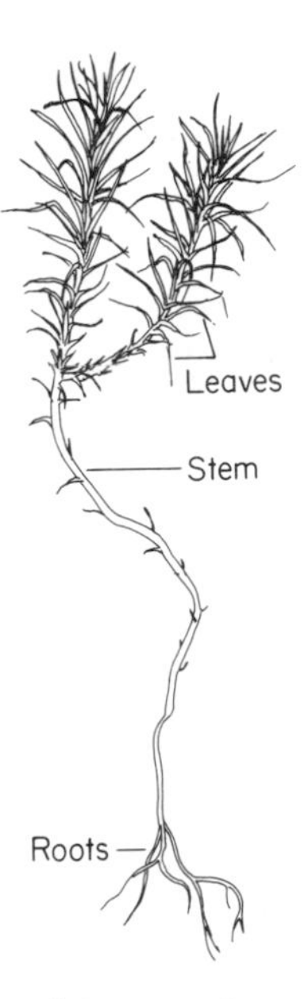

Figure 94. Vegetative organs of *Lycopodium* (modified after Foster and Gifford, 1974).

between roots and underground stems. The stem bears buds that can develop leaves. The apices of roots do not have this ability. Furthermore, the apical meristem of a root is protected by a root cap. There is no comparable structure on the apices of stems. In vascular plants that have true roots the sporophyte embryo gives rise, at an early stage, to a primary root and a primary leafy stem. Thus roots and stems are developmentally differentiated structures. The stems of many species of *Selaginella*, a modern lycopod, have xylem vessels as well as tracheid tracheary elements. This genus has existed for at least 350 million years. It seems highly likely that all of the xylem tracheary cell types now known evolved within the first 50 million years of life on land.

Leaves of clubmosses and other vascular plants arise as lateral protuberances from a stem apex (bud). A leaf is a determinate organ in contrast to the open or theoretically unlimited type of growth of the stem or root.

The leaves of clubmosses are called **microphylls**, an unfortunate name as some may reach lengths of 50 cm or more. A microphyll is distinguished from a **megaphyll** (found in ferns, gymnosperms and angiosperms) by its very simple vascular system. The vascular supply of a microphyll consists of a single strand that emerges from the vascular tissue of the stem and extends as an unbranched mid vein through the leaf. Megaphylls have complex venation patterns and the xylem strand is continuous with the peripheral xylem of the stem. Immediately above the point at which the xylem strand of a megaphyll separates from the stem xylem there is a break in the stem vascular tissue called the **leaf gap**, and this gap is filled with non-vascular parenchyma cells (Fig. 95).

Some of the living lycopods have life histories that are similar to those of *Psilotum* and *Tmesipteris* (Psilopsida). A large sporophyte alternates with a free-living gametophyte that bears archegonia and antheridia. In the simplest clubmosses the sporangia develop on the upper surfaces, or in the axils, of ordinary leaves. In more complex forms the sporophylls are differentiated in size, shape and color from ordinary cells. The collection of differentiated sporophylls is a strobilus or cone. The sporangia have protective walls, the innermost layers of which produce nutritive substances for the developing spores. When the spores are mature the sporangia split open and the repro-

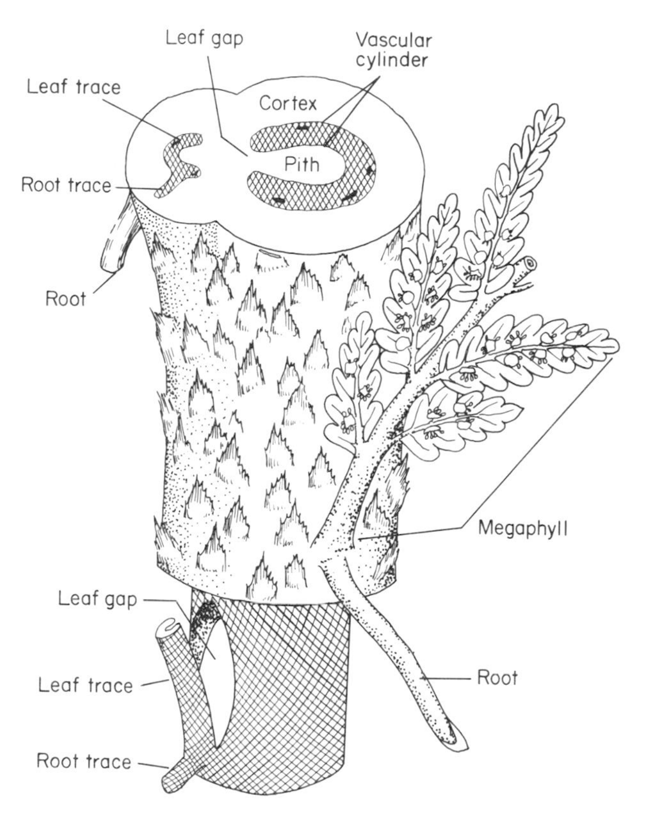

Figure 95. Block diagram showing the leaf gaps of a fern megaphyll (modified after Jeffrey, 1917).

ductive cells are released. Spores develop into subterranean or subaerial gametophytes that produce motile sperms. The eggs are protected within archegonia. The sperms must swim through a film of water in order to effect fertilization.

The type of life history described above is homosporous. In many lycopods the life history is heterosporous (Fig. 96). Both male and female gametophytes are very reduced in heterosporous forms. The male gametophyte develops entirely within the spore wall which does not rupture until the sperms are released. The female gametophyte may have chlorophyll in its cells, but the amount of photosynthate produced is small in relation to the amount of food stored within the female gametophyte before the megaspore is released from its sporangium. As the female gametophyte develops, the spore wall breaks open and archegonia form on the exposed surface (Fig. 97).

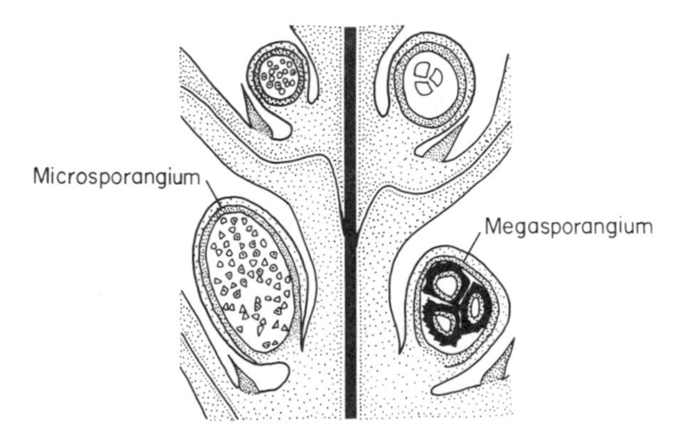

Figure 96. Longitudinal section through microsporangia and megasporangia of a fossil *Selaginella* (modified after Stewart, 1983).

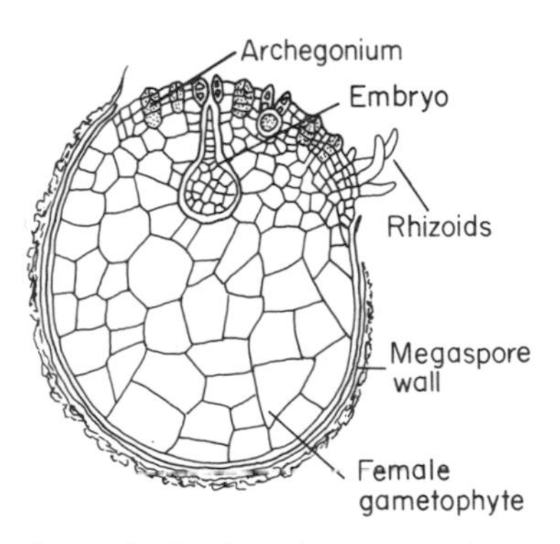

Figure 97. Section through the female gametophyte of *Selaginella* (modified after Bruchmann, 1912).

The modern lycopods represent only a fraction of the diversity that has existed in the group. About 300-345 million years ago lycopods formed the dominant vegetation on land. Some forms developed the tree habit and grew to heights of >50 m (Fig. 98). These giant forms developed a cambium meristem in the trunk and produced secondary xylem tissue. Surprisingly little wood was formed by this cambium. It appears that parenchyma cells outside the vascular tissue assumed a meristematic function. The bulk of the trunk tissue of one of these trees consisted of cortical parenchyma and this must have had a mechanical support role.

Class Sphenopsida

Only one genus of this class (*Equisetum*) has survived into modern times. *Equisetum* is very ancient. The genus is probably 300 million years old. During the period when the tree lycopods were dominant the sphenopsids also flourished. Both herbaceous and tree forms were abundant. *Equisetum*, the horsetail, is herbaceous.

The sphenopsids have much in common with the lycopods. The

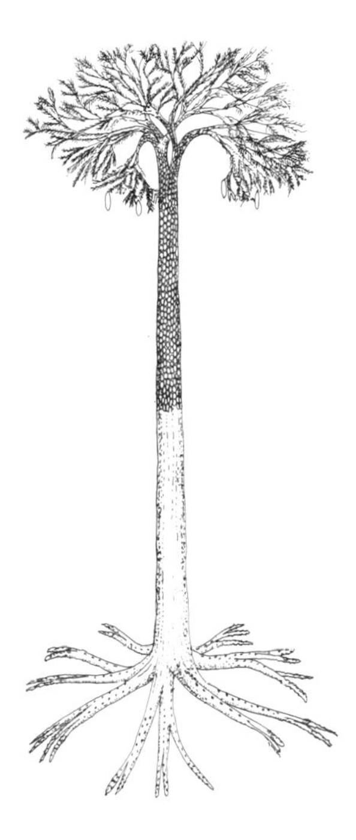

Figure 98. *Lepidodendron*, a coal age clubmoss that grew as a tree (modified after Stewart, 1983).

sporophytes have roots and microphyllous leaves (one extinct order had megaphylls). Modern forms are homosporous. Heterosporous forms are now extinct. The most distinctive feature of the sporophyte is the peculiar jointed nature of the stem which is divided into nodes, which bear whorls of leaves, and internodes which do not bear appendages (Fig. 99).

The tree-like sphenopsids of the coal age (~330 million years ago) also had the peculiar jointed stems of *Equisetum* (Fig. 100). A massive trunk of secondary xylem developed from the meristematic cambium.

Class Filicopsida

The members of this class are called ferns and are distinguished from the other classes of vascular plants considered above (except for one extinct order of Sphenopsida) by the possession of megaphyll leaves. The stems on which the leaves are borne often have complex vascular cylinders. True roots are present.

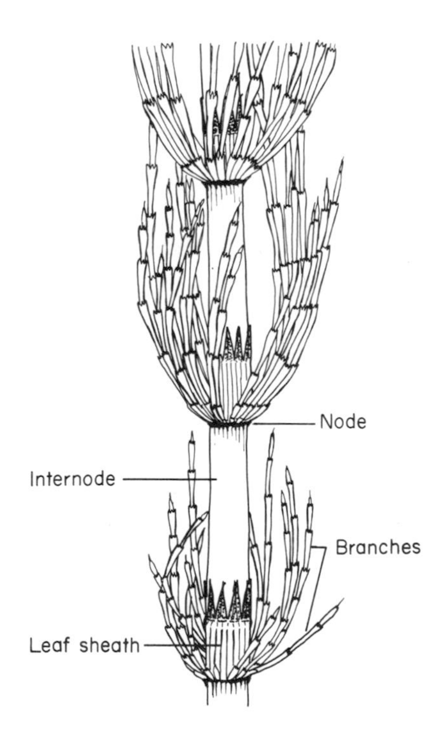

Figure 99. Portion of the aerial axis of *Equisetum* showing nodes, internodes leaf sheaths and branches.

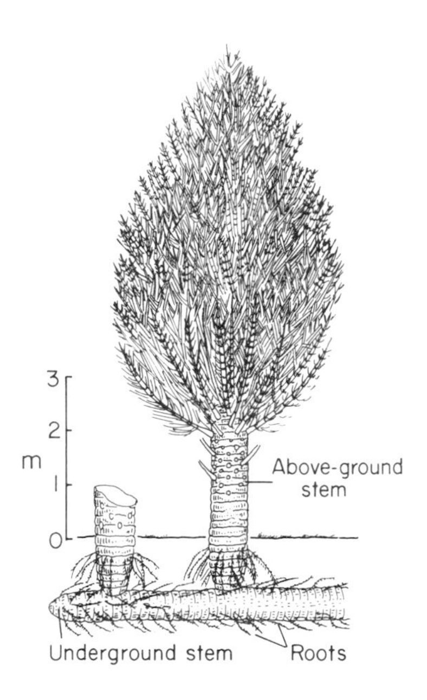

Figure 100. *Calamites*, a coal age sphenopsid that grew as a tree (modified after Emberger, 1968).

The ferns are mostly homosporous with free-living gametophytes. Some are heterosporous and the gametophytes develop within the spore walls. Most ferns have sporangia that are borne on the undersurfaces of leaves. The sporophylls may be differentiated and in cones, or they may be undifferentiated.

The most conspicuous feature of ferns is the development of massive leaves (Fig. 101). It is generally believed that megaphylls evolved by the **webbing** of complete branch systems (Fig. 102) after **flattening** in a single plane. There is a huge diversity of megaphylls among the ferns.

The Filicopsida is a very ancient class with a fossil record extending back nearly 350 million years. This is a very successful group of vascular plants – there are more than 10,000 modern species. There was extensive diversification of now extinct groups through the coal age (280-345 million years ago). Tree-like ferns reached heights of ~8 m. The massive trunks developed in an interesting way. Roots were initiated high up on the stems and these grew down through, or on, the surface of the trunk to ground level. This mantle of roots provided mechanical support for the tall stems.

The huge diversity of modern fern groups had its origins in the Jurassic era of 141-195 million years ago. Most extant ferns are herbaceous and are dwarfed by the tree forms of modern seed plants.

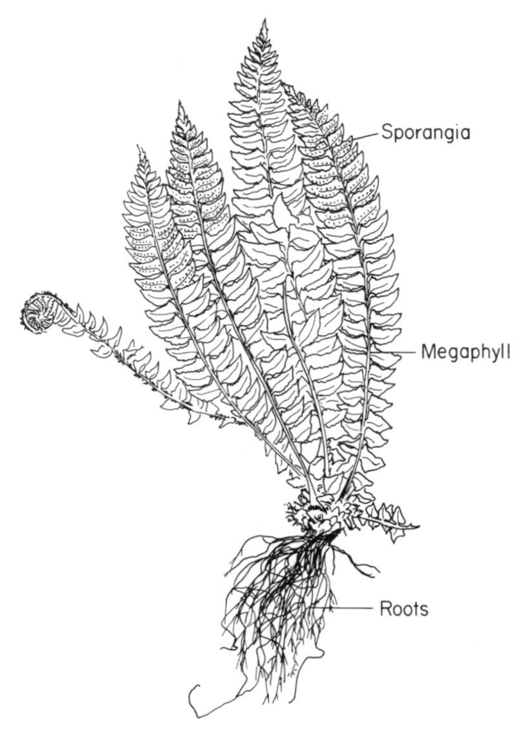

Figure 101. Sporophyte of *Polystichum* (Filicopsida) with megaphylls (modified after Stewart, 1983).

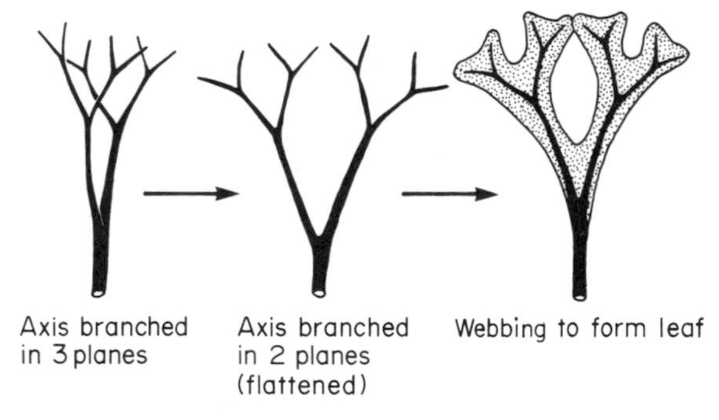

Figure 102. Theoretical sequence of megaphyll evolution from a system of axes (modified after Stewart, 1983).

Nearly all ferns lack a vascular cambium, so secondary wood formation cannot occur. This may explain the diminutive sizes of most members of the group. Although secondary xylem is rare, the arrangement of primary xylem in the stem is often quite complex. In the simplest forms there is a simple axile rod of vascular tissue. In more advanced forms the xylem is arranged in a dissected peripheral cylinder. Sclerenchyma development is extensive in ferns.

Class Progymnospermopsida

This class of vascular plants is now entirely extinct. Fossils show that progymnosperms were present between about 320 and 350 million years ago. The group may have evolved from the Trimerophytopsida along with the ferns. Progymnosperms are believed to be the ancestors of all seed plants.

Members of the class were heterosporous and free sporing (i.e., the spores were all shed), hence their reproductive characteristics were very similar to those of ferns. However, the anatomy of progymnosperms was much more similar to that of modern gymnosperms. Specifically, the cambium was bifacial, that is, secondary xylem was produced from its inner face and secondary phloem from its outer face. The cambia of other tree-like forms described above produced no secondary phloem. In addition, the secondary xylem of progymnosperms was distinguished from other contemporary vascular plants by the presence of **wood rays**. Wood rays are tracheary elements that conduct solutes radially. Rays have a radial orientation, contrasting with the predominantly longitudinal disposition of tracheary elements (Fig. 103).

The progymnosperms and other vascular plants of 350 million

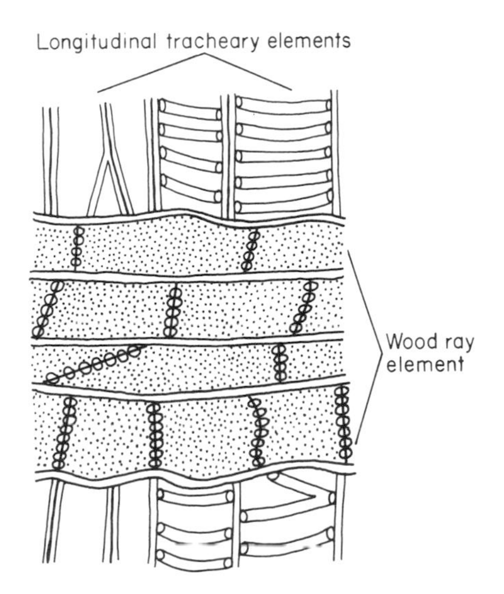

Figure 103. Secondary xylem of seed plant showing wood rays (modified after an original by K. E. von Maltzahn).

years ago shared between them nearly all the anatomical developments found in modern land plants. Within a space of 30-50 million years of the first colonization of land the major problems of terrestrial existence had been solved by plants.

Class Gymnospermopsida

Trees of the great coniferous forests of the world are gymnosperms. Clearly, members of this class represent a major component of the terrestrial flora. By 250 million years ago the gymnosperms were the dominant plants on land, and this dominance continued for perhaps 180 million years.

The Gymnospermopsida is distinguished from all the lower vascular plants by the seed habit and by the transmission of male gametes through pollen tubes. The gymnosperms are archegoniate, in contrast with the angiosperms which lack gametangia. The seeds of gymnosperms are naked, whereas angiosperm seeds are enclosed within a carpel.

Modern gymnosperms are broadly divided into two groups, the coniferophytes and the cycadophytes. Coniferophytes are familiar to all residents of cool/temperate regions of the earth. Spruce, fir, juniper, pine and redwood are all well-known conifers. These trees all have profusely branched trunks and small needle leaves. Massive secondary xylem wood development is common and forms the basis of the largest forestry industries in the world.

Cycadophytes are much less familiar plants. To the untrained eye they resemble palm trees. The leaves are large and pinnate, like those of ferns. The stems are little branched and bear a terminal crown of large leaves, giving the palm-like appearance. Very little secondary xylem is formed in the wood of these plants.

Most of the cycadophytes are now extinct. Those which survive have what must be regarded as an evolutionary relic in their reproductive system. They have flagellate sperms. A pollen tube develops from the male gametophyte, as in all gymnosperms. Unlike coniferophytes, all cycadophytes inject motile sperm cells through the pollen tube and into a small chamber above the female gametophyte. Fertilization is internal, but involves flagellate sperm. In all other seed plants the male gamete is non-motile.

In gymnosperms and all seed plants the megasporangium does not shed its spores. Instead, the female gametophyte develops within a protective jacket on the parent sporophyte. The only mechanism that can provide access for sperms to eggs enclosed in this way is the pollen tube. It seems, therefore, that the seed habit and the pollen tube evolved together. A seed is formed after an embryo forms from the fertilized egg cell.

When did the seed habit originate? The answer is a little complicated, but it seems certain that by 350 million years ago integumented megasporangia had evolved among the vascular plants. A fully enclosed megasporangium with micropylar access had certainly appeared in the lower coal age (325-345 million years ago). Pollen grains have also been found in coal age deposits.

Within 50-60 million years of the first appearance of *Rhynia*, plants had found solutions to all the terrestrial problems listed in Chapter 5. It was a truly remarkable progression from simple green algae (which may have been unicellular) to tree-like plants with mechanical and conducting tissue, cuticles, stomata, roots, leaves, seeds and pollen grains.

Class Gnetopsida

This is a small class of seed plants with unknown evolutionary relationships. There are only three living genera, *Ephedra*, *Gnetum* and *Welwitschia* .

Gnetopsids have more in common with gymnosperms than other vascular plants. Both groups contain archegoniate seed plants. However, gnetopsids stand apart in the details of their wood anatomy and reproductive organization. Xylem vessels are lacking in the gymnosperms, but are a conspicuous feature of the wood of gnetopsids. It seems likely that vessels evolved independently in gnetopsids.

The sporangial cones of gnetopsids are complicated compound structures (Fig. 104). The sporangium-bearing shoots are arranged

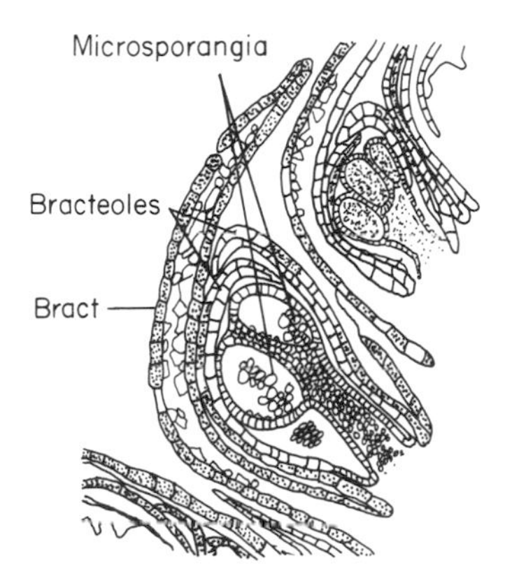

Figure 104. Longitudinal section of a microsporangial cone of *Ephedra*.

among sterile leaf-like structures, the bracts. In *Ephedra* the cone bearing megasporangia consist mostly of sterile bracts. The cones of most species of the genus bear only two megasporangia in the axils of the uppermost bracts. The intermix of sterile and fertile leafy structures in the cones of gnetopsids has often been compared to that in flowering plants. The megasporangium of gnetopsids is enclosed by an integument, as in all gymnosperms, but there is, in addition, another outer envelope. The integument and outer envelope form a long micropylar tube (Fig. 105). This organization is absent from the gymnosperms.

Gnetum and *Welwitschia* do not develop archegonia, whereas gametangia are conspicuous in the megagametophytes of *Ephedra*. In *Welwitschia* fertilization occurs in the pollen tube. This is unique among seed plants.

Class Angiospermopsida

The members of this class are called flowering plants. The only diagnostic feature of the group is double fertilization and development of a polyploid endosperm (Chapter 9).

There are certain general features shared by most, but not all,

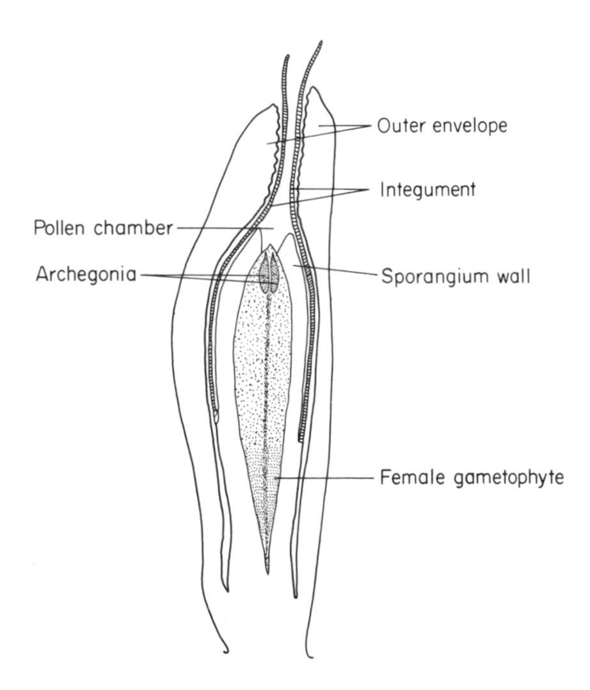

Figure 105. Longitudinal section of the female gametophyte of *Ephedra* (modified after Maheshwari, 1935).

flowering plants. The flower is an obvious characteristic and this normally bears sterile accessory structures like petals and sepals. In most angiosperms the megasporangium is fully enclosed within two integuments. The female gametophyte is very reduced (usually seven cells and two polar nuclei) and without archegonia. The stamens are simple, consisting of filament and anther. Most angiosperms have xylem vessels and phloem sieve elements with companion cells. This combination of reproductive and anatomical characteristics describes most angiosperms. There are exceptions. Apart from double fertilization and endosperm formation, each of the traits listed above may be found among gymnosperms or gnetopsids. Since flowering plants are thought to have evolved from gymnosperms, this is hardly surprising.

The angiosperms are the dominant component of the modern terrestrial flora. The class became dominant about 130 million years ago and since that time a huge number of species has evolved (at least 200,000). The morphological and reproductive diversity is overwhelming, and no attempt can be made here to present a comprehensive review.

The diversity of angiosperms is classified into two large subclasses, the Dicotyledoneae and the Monocotyledoneae, known as the dicotyledons and monocotyledons, respectively. The dicotyledons have two embryo leaves. The monocotyledons have one. Further distinctions are as follows:

(a) The parts of the flowers of monocotyledons occur in threes or in multiples of three, whereas in dicotyledons the parts occur in fours or fives,

(b) Monocotyledons have parallel leaf veins, contrasting with the reticulate veins of dicotyledons,

(c) The vascular bundles are scattered through cross-sections of the stems of monocotyledons, whereas in dicotyledons the bundles occur in cylinders,

(d) Dicotyledons have a bifacial cambium that is absent from monocotyledons.

The evolutionary origins of the angiosperms are not apparent from the fossil record. The group appeared suddenly about 136 million years ago at a time when mass extinctions were occurring among plant and animal groups. Many of the gymnosperm groups disappeared, only the conifers persisted in numbers. About 25% of the known families of animals disappeared at that time (including the dinosaurs). The reasons for these changes in the earth's flora and fauna may have been climatic, but this is speculative.

11

Bryophytes

Distinguishing Characteristics

Bryophytes (mosses, liverworts and hornworts [Table 2]) are almost all terrestrial plants. They have the same photosynthetic pigments as members of the Tracheophyta and Chlorophyta (green algae). Bryophytes are distinguished from green algae by the possession of archegonia and antheridia. Furthermore, bryophytes have water conducting cells that are not found among the seaweed divisions. Bryophytes are protected by a thin epidermal cuticle.

The bryophytes are distinguished from vascular plants in several important ways that are related to water supply on land. First of all, the elaborate regulatory stomatal mechanism of tracheophytes is absent from the dominant gametophytes of bryophytes (some bryophyte sporophytes have stomata). Second, roots (found in all extant vascular plants but the psilopsids) are absent in bryophytes. Third, the substance lignin that is characteristic of xylem tracheary elements has not been reliably reported from bryophytes.

The life histories of bryophytes are fundamentally similar to those of vascular plants that have archegonia (Fig. 106). However, the development of the sporophyte is very limited in bryophytes and there is no differentiation into stems, leaves and roots. The bryophyte sporophyte has no direct connection with the soil and must depend

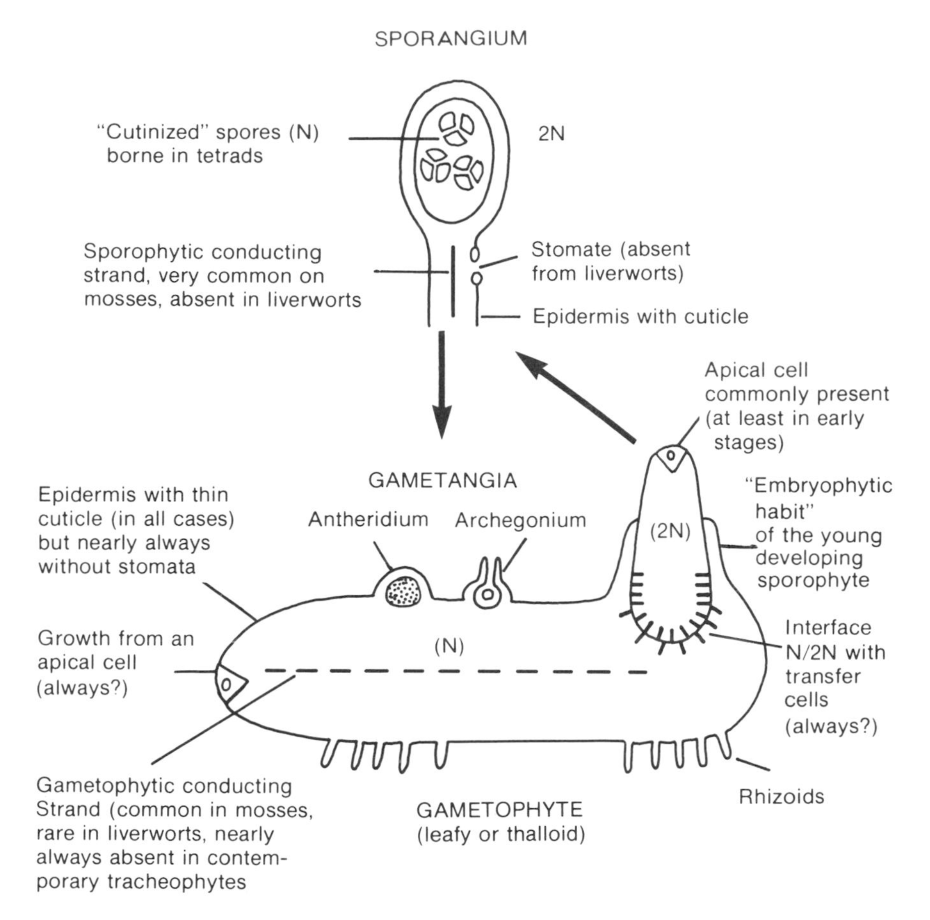

Figure 106. Stylized diagram showing major life history features of bryophytes (modified after Hébant, 1977).

on the gametophyte (to which it is attached) for access to soil water and minerals.

The gametophyte generation of bryophytes is more conspicuous and longer lived than the sporophyte. The gametophyte may become a highly differentiated leafy structure, quite unlike the gametophytes of vascular plants. In spite of the high degree of differentiation, most bryophyte gametophytes are small (<50 cm tall) and restricted to moist habitats.

Bryophytes have a water economy that is fundamentally different from that of vascular plants. Whereas vascular plants are homiohydric, all bryophytes are poikilohydric. Elaborate drought avoidance mechanisms are absent in bryophytes, which must, on occasion, endure reduced tissue water content.

Gametophyte Functional Forms

Ten major functional forms of terrestrial bryophyte gametophytes have been distinguished. The occurrence of each of these forms is closely correlated with water availability. Examples of the ten life forms are shown in Fig. 107. In bogland and moist tundra the life form of **tall turfs** predominates. These plants have access to considerable quantities of soil water and the crowded shoots, dense foliage and well developed rhizoid system have considerable capacity for water conduction.

In **mat**, **weft**, **tail** and **fan** forms external capillary water conduction is unimportant, whereas capillary retention of water is considerable. After rainfall the retained water allows a continuation of metabolic activity during a subsequent dry period. Some bryophyte mats can hold 5-10 times their dry weight of water. Short **turf** and **cushion** forms growing on exposed surfaces can also retain considerable quantities of water. **Pendant** forms are limited in their distribution to tropical cloud forests. The growing tip of the hanging plant is continually bathed in falling water and this allows growth to continue without interruption (leading to the characteristic axial elongation).

Additional correlations have been made between the life forms of bryophytes and the habitats in which they are found. However, it should be pointed out that these correlations have not been subjected to rigorous experimental analyses. The subject is not pursued further here.

Water Relations of Individual Gametophytes

Bryophytes have a greater diversity of water uptake and transport mechanisms than vascular plants. Two groups of mechanisms may be distinguished. Plants with the **endohydric** mechanism take up water from the soil and conduct it to the evaporating photosynthetic surfaces. **Ectohydric** bryophytes absorb water (and lose it) over their entire surface. These plants are often found on bare rock where atmospheric moisture or run-off is the only source of water.

In most endohydric forms water is taken up from the soil by rhizoids (Fig. 106). The rhizoids of mosses are multicellular and usually brown in color. In some species the rhizoids are produced over the whole length of the stem. Liverworts and hornworts have unicellular rhizoids. The rhizoids act as an inert felt, like an absorbent layer of paper toweling between the plant and the substratum. The exact role of rhizoids is not fully understood, but it appears that

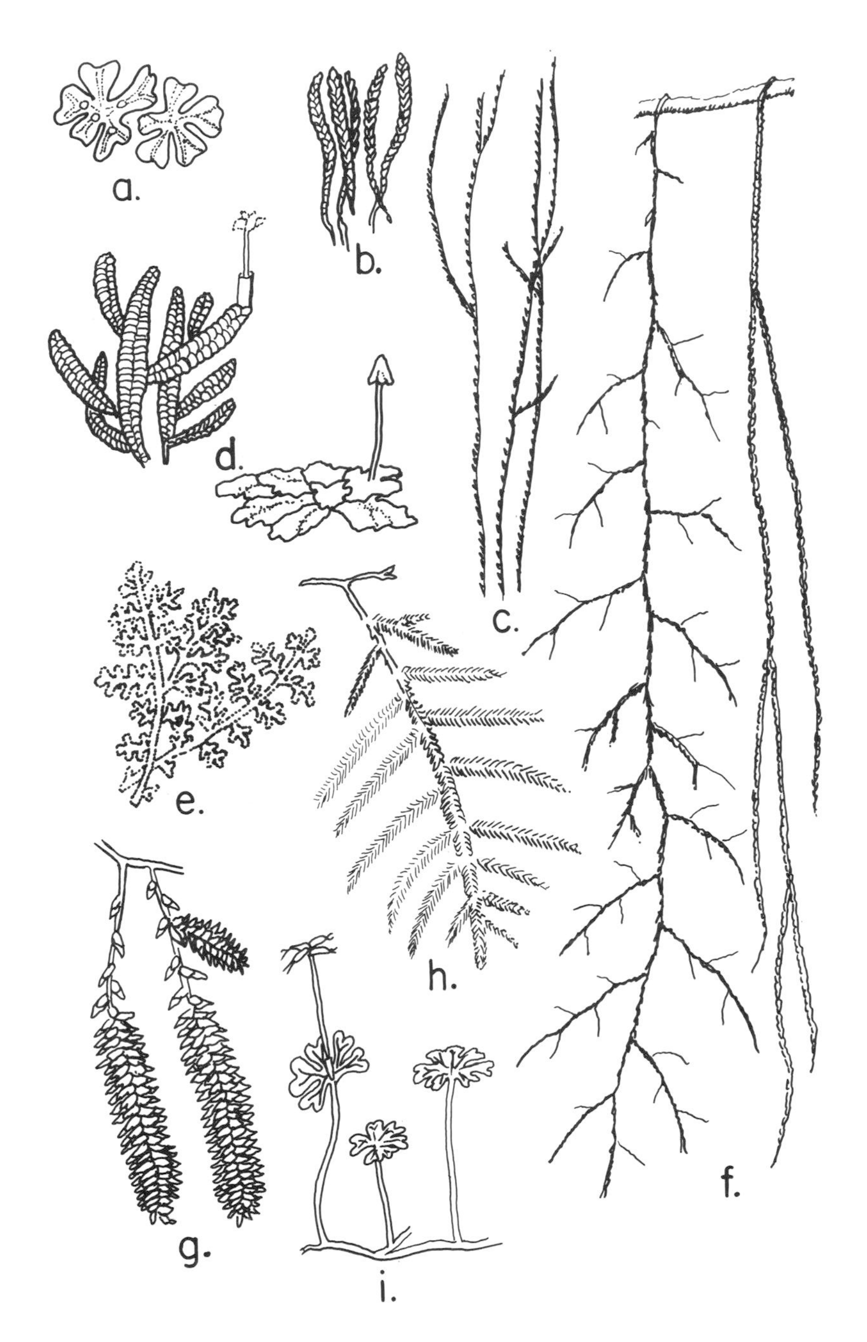

Figure 107. Life forms of liverworts (an equivalent series occurs in mosses): A. Annual; B. Short turf; C. Tall turf; D. Mat; E. Weft; F. Pendants; G. Tail; H. Fan; I. Dendroid (modified after Mägdefrau, 1982, by permission of Chapman & Hall Ltd.).

water uptake in bryophytes is due to the swelling of cell walls, mucilages, etc. that are external to the cell membranes. The analogy with absorbent paper toweling seems appropriate.

Within the gametophyte tissues of endohydric bryophytes water is moved through the following pathways:

(a) Through specialized conducting cells called **hydroids**,

(b) Through the free space of cells walls,

(c) Cell to cell through intervening walls and cell membranes.

The driving force through all these pathways is the water potential gradient between soil, plant and atmosphere.

Hydroid development is very limited in liverwort gametophytes. In the larger mosses most water conduction occurs through this pathway. Hydroids are dead, elongated cells (Fig. 108). The walls are often thin and highly permeable to water. This combination of characteristics makes the hydroid strand a preferential pathway of low resistance to water movement. Hydroids function in the same way as

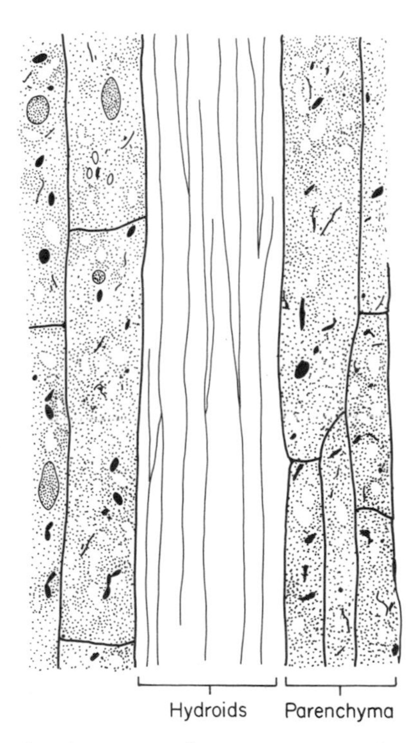

Figure 108. Longitudinal section of a moss gametophyte stem showing the elongate hydroids (modified after Hébant, 1977).

xylem tracheary elements of vascular plants. However, lignified cell wall thickenings are absent from bryophytes.

In liverworts (and some mosses) most internal water conduction is through the free space of parenchyma cell walls. Cell walls ($\sim 1 \mu m$ thick) have a much greater (100 times) conductivity than cell membranes. This pathway must be especially important in bryophytes with small, thick-walled cells. In gametophytes that have large, thin-walled cells water conduction must occur largely from cell to cell through cell membranes. Water flow is ultimately driven by evaporation from the plant into the atmosphere. Although many bryophyte gametophytes have a cuticle, the leaf resistance to water vapor diffusion is generally low.

Some gametophytes have surface pores and associated internal air chambers (Fig. 109). The pores do not regulate the flow of gases in or out of the air chambers. Since there are no guard cells, no variable resistance a water flow regulation mechanism exists. The pores stay wide open during severe drought.

Even bryophytes with differentiated internal conducting systems transport water externally. In the moss *Polytrichum*, which has a large conducting strand, internal transport alone maintains high turgidity when there is 90% atmospheric relative humidity. At 70% relative humidity both internal and external conduction are required to retain turgor. In bryophyte gametophytes without conducting strands nearly all transport takes place externally.

External conduction takes place in capillary channels that are 10-100 μm across.There is a large diversity of systems involving spaces between overlapping leaves, between stems and sheathing leaf bases, etc. (Fig. 110).

Bryophyte gametophytes do not have well developed drought avoidance mechanisms like those of vascular plants. For this reason bryophytes are poikilohydric and must tolerate massive reductions in water content. Under wet conditions the tissues may contain 90% water by weight. Under drought the tissue water content is reduced to 5% or less. The mechanisms of drought tolerance are unknown. During reduced water content of the tissues, cellular metabolism is very reduced. Recovery occurs following wetting.

The variable water content of bryophytes can be partitioned into **apoplast** water (held in cell walls), **symplast** water (held within the protoplasm) and external **capillary** water. The change in symplast water content occurs between water potentials of about -5 to -200 bars. Between about -1 and -5 bars water potential, most of the change is in the water associated with the cell walls. Above -1 bar water potential, most of the change in water content is in the external

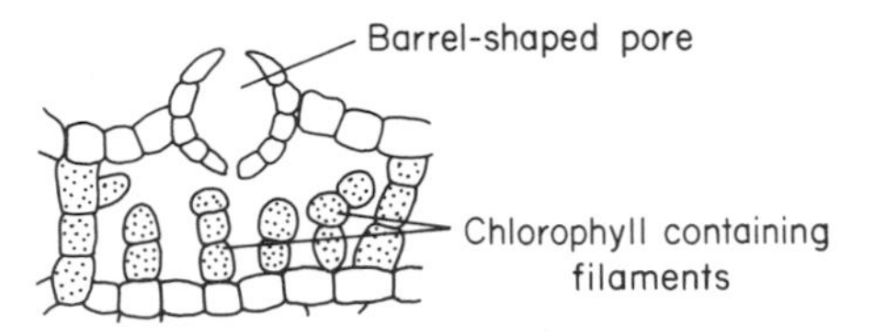

Figure 109. Gas exchange pore and internal air space of *Marchantia*, a thallose liverwort (modified after Parihar, 1961).

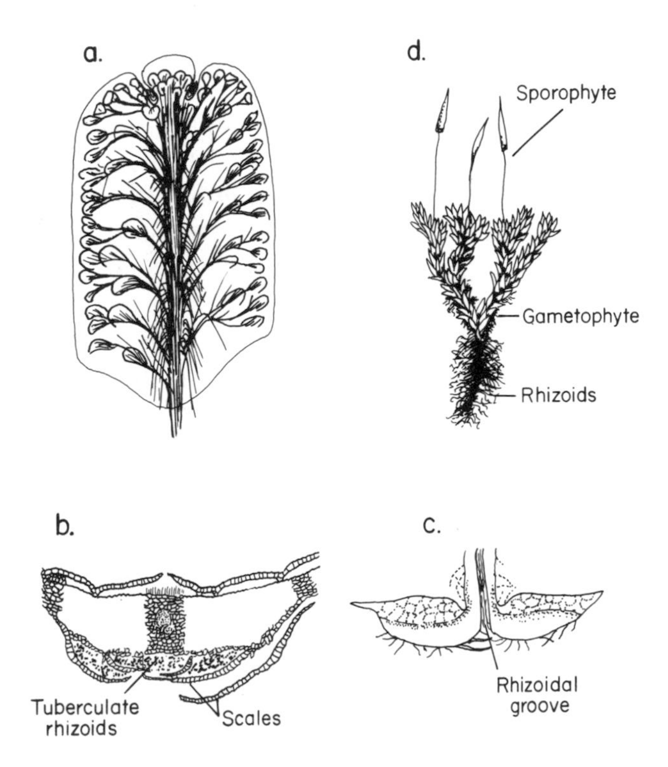

Figure 110. Structures involved in external capillary conduction of water. A. *Marchantia* lower surface showing scales and rhizoids that facilitate conduction; B. Section through *Conocephalum* (thallose liverwort) showing scales and rhizoids; C. Section through the base of the archegoniophore of *Fimbriaria* (liverwort) showing rhizoidal groove; D. Leafy moss with dense rhizoids and leaves that aid in capillary conduction (modified after Hébant, 1977).

capillary water. In ectohydric species from dry habitats apoplast water and external capillary water form the largest proportions of total water content. A large part of a gametophyte's water can thus be lost before the symplast water is reduced and turgor is lost.

Water Relations of the Sporophyte

The sporophyte of mosses and liverworts consists of a **foot** which is buried within gametophyte tissue, a **seta** (stem) and a **capsule** which bears the spores (Fig. 106). The vegetative morphology of the sporophyte is much less variable than that of the gametophyte. In addition, the life span of the sporophyte is much shorter than that of the gametophyte.

The foot of the bryophyte sporophyte has no direct access to soil moisture. Water is supplied internally and by external capillary conduction. The foot appears to establish connection with the supporting gametophyte through special **transfer** cells (Fig. 111). The cell wall convolution of these cells greatly increases the surface area of the underlying cell membranes. There is, however, no direct cytoplasmic connection between gametophyte and sporophyte generations.

Within the setae of mosses there is often a very well-developed conducting strand of hydroids. Liverwort sporophytes have no special conducting tissues and must rely on water transfer through the free space of cell walls and through parenchyma cytoplasm. The conducting strand of the moss seta ends at the base of the capsule. Further water transfer is through ordinary parenchyma cells.

The sporophytes of mosses and hornworts have the only functional stomata found in the division Bryophyta. The guard cells open and close the pores that they surround and respond to CO_2 and water vapor concentrations in the same way as vascular plant guard cells.

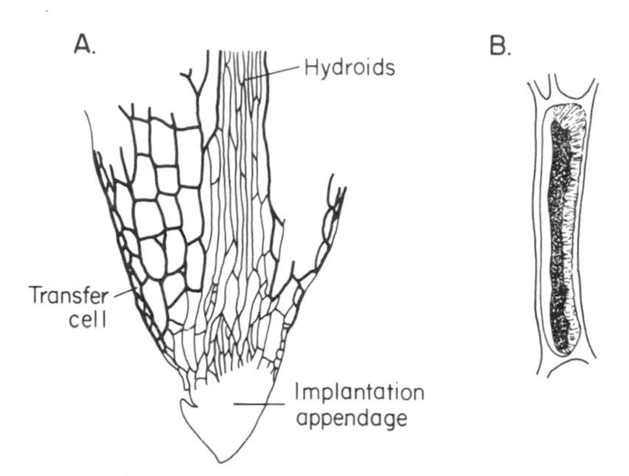

Figure 111. Longitudinal section through the base of the sporophyte of *Polytrichum* (moss) showing foot with transfer cells (A.) and enlarged illustration of a transfer cell (B.) showing cell wall ingrowths (modified after Hébant, 1977).

Transport of Photosynthate

Many mosses (but not liverworts) have specialized cells called **leptoids** for the transport of photosynthate about the plant body. Leptoids bear a strong resemblance to the sieve cells of vascular plants. They have slightly oblique end walls with pores, and degenerated nuclei when mature (Fig. 112). Like vascular plants, the major translocate in mosses is sucrose. The maximum transport velocity is about 32 cm/hr.

Mechanical Cells

Although liverworts are rather small in comparison with vascular plants, specialized mechanical support cells (**stereids**) do develop. These cells have thickened lateral walls and are alive when mature (Fig. 113). Further support is obtained by swelling of parenchyma cell walls. The inner layer of the wall is rich in hemicelluloses which take up water more readily than the outer cell wall layers. This creates a swelling pressure of the inner wall layer on the outermost layers. The hydraulic pressure created is functionally equivalent to the mechanical support obtained by turgor pressure.

Sexual Reproduction

The antheridia and archegonia of bryophytes bear a striking resemblance to those of vascular plants (Fig. 114). The developing gametes

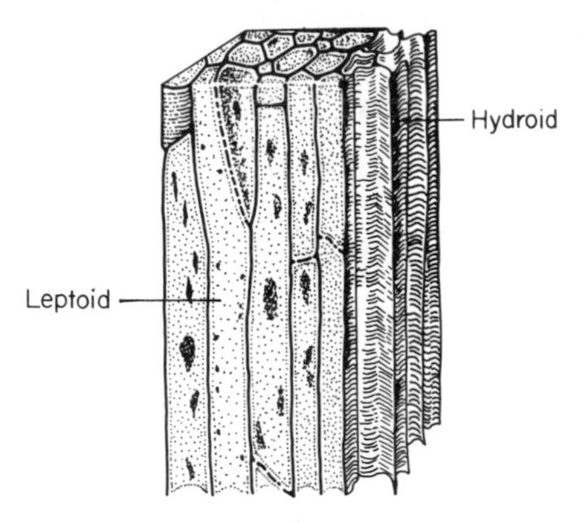

Figure 112. Block diagram through the conducting tissues of *Polytrichum* (moss) showing leptoids and hydroids (modified after Hébant, 1977).

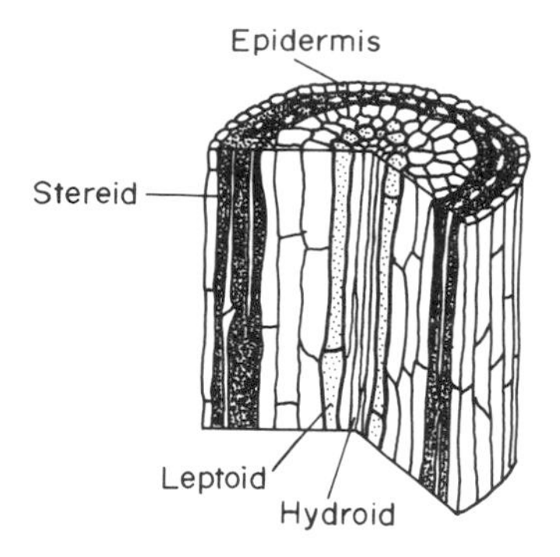

Figure 113. Block diagram showing internal structure of the sporophyte seta of *Funaria* (moss). Notice the stereids (mechanical cells)(modified after Hébant, 1977).

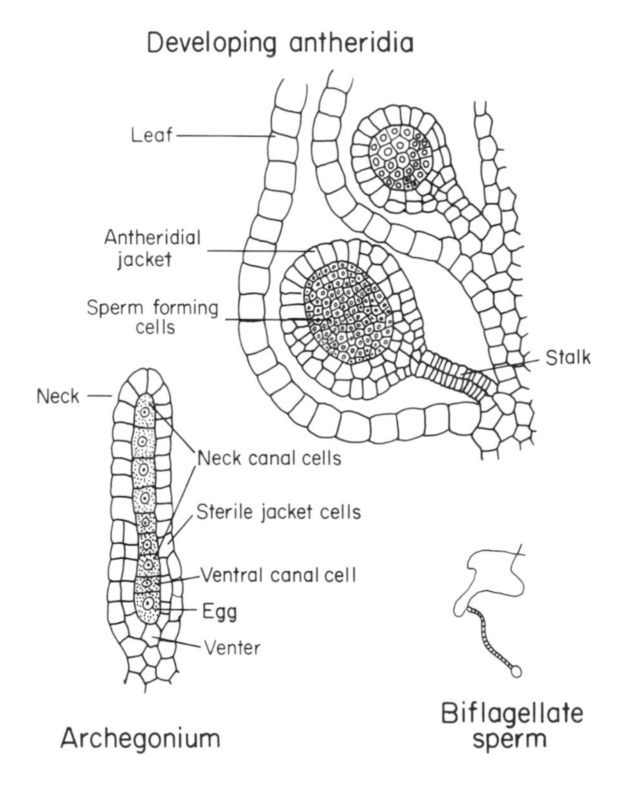

Figure 114. Developing gametangia and sperms of a leafy liverwort. The neck canal cells of the archegonium disintegrate leaving a passage for sperm access (modified after Parihar, 1961).

are always protected by sterile jacket cells. The gametangia themselves are often protected by pits or enclosing leaves of the gametophyte.

In some of the liverworts the archegonia and antheridia are borne on special elevated axes called **archegoniophores** (Fig. 115) and **antheridiophores**, respectively. Fertilization occurs before the axes elongate so the sperms do not have to climb the archegoniophore stalks. The sperms of bryophytes are elongate and biflagellate. The sperms of each antheridium are extruded as a mass under moist conditions. If contact is made with the air-water interface the whole mass disperses immediately. A monolayer of sperm mother cells forms on the surface of the water and individual sperms free themselves from the shells of mucilage that surround them.

Although liberated sperms of bryophytes remain motile for up to 6 hours, they seem very reluctant to swim far during that time. Sperms released in a pool of water remain crowded near the point of liberation. Transfer to the archegonia is achieved in ways other than flagellar swimming. Some mosses have splash cup mechanisms (Fig. 116). The antheridia are surrounded by a cup of stiff leaves. Raindrops falling in the cups carry away sperms. The leaves of the archegonium-bearing plants point up so that water drops (which may carry sperms) are caught. The water runs down between the leaves to

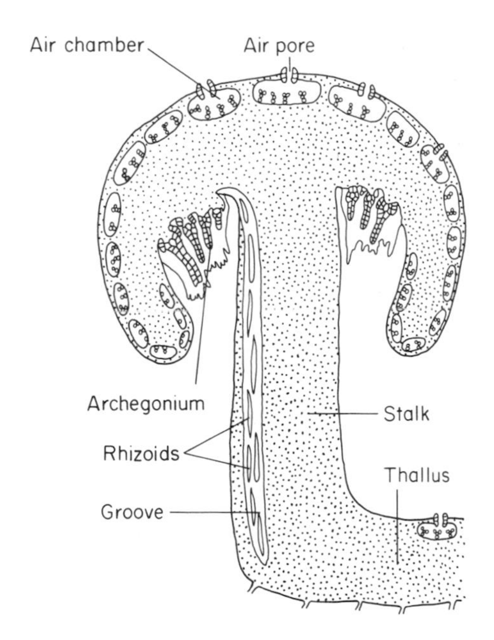

Figure 115. Archegoniophore of *Marchantia* (liverwort) (modified after Parihar, 1961).

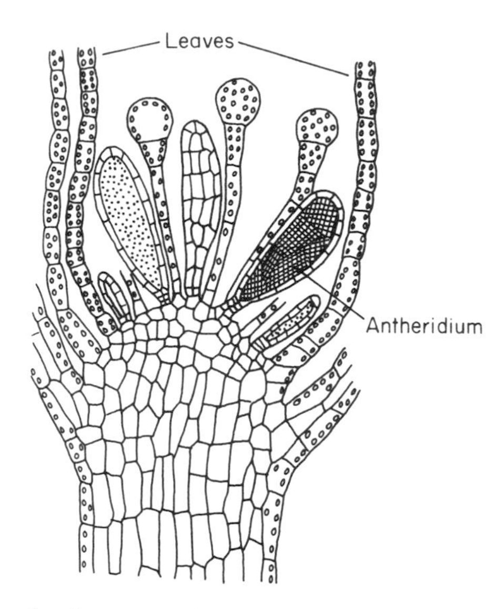

Figure 116. Longitudinal section through the splash cup of *Funaria* (moss) showing ring of leaves around developing antheridia (modified after Smith, 1955, by permission).

the archegonia. In large mosses sperms may be carried 1-2 m in this way.

In other bryophytes sperm transfer seems to take place passively through water movement during rain showers or by movement of water films when the lipids discharged with sperm masses reduce surface tension. Whatever the details, sperm transfer in all bryophytes is completely dependent on water.

Sperm cells arriving in the vicinity of an archegonium are strongly attracted by archegonial exudate. The attractant seems to be sucrose. The sperm moves down the archegonial neck and effects fertilization.

Sporophytes and Spore Production

The zygotes produced by fertilization develop into a diploid sporophyte generation. The diversity of form among bryophyte sporophytes is limited in comparison with the gametophytes that support them. All sporophytes have a foot embedded in a gametophyte and a capsule in which spore development occurs. In mosses and liverworts the foot and capsule are separated by a stalk, the seta. Growth is apical in these two groups (at least in the early stages). The hornworts, in contrast, have no seta and growth occurs in the transition zone between capsule and foot. The sporophyte capsules of the three classes of bryophytes (mosses, liverworts and hornworts) are quite divergent in their structure and function. They are best considered separately.

The liverwort capsule has a sterile jacket of one or more cell layers (Fig. 117). Inside the jacket there may be spores alone, but in most genera the spores are mixed with sterile cells that have spiral wall thickenings (Fig. 117). These sterile cells are called **elaters**. Elaters function in spore dispersal after the mature capsule splits open. After exposure to the atmosphere the elaters begin to dry. As they do the spiral wall thickenings twist and put the cell water under tension. Further drying produces a tension which breaks the cohesion of water molecules to the internal cell walls. The spiral wall thickenings immediately spring back to their original shape throwing spores into the air.

The capsules in mosses are often relatively complex structures. In the simplest forms the capsule does not have specialized dispersal

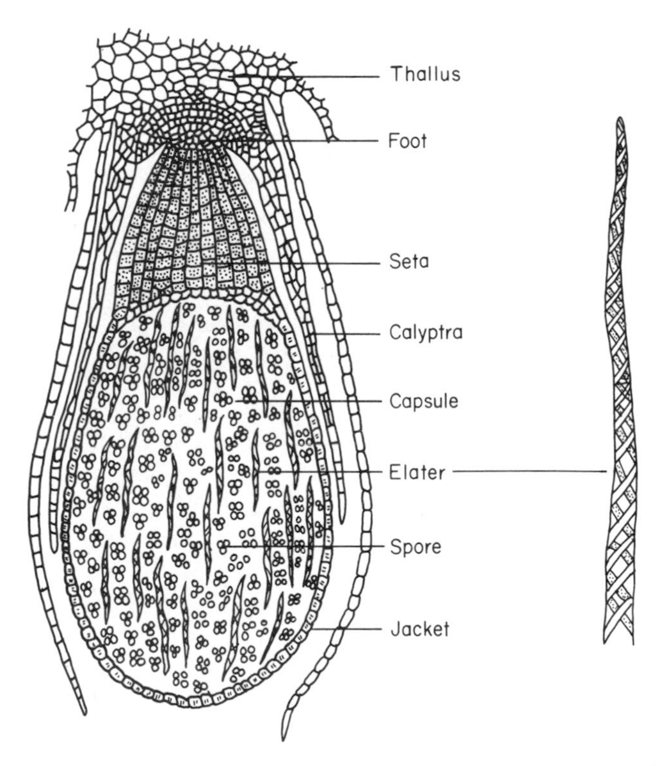

Figure 117. Longitudinal section through the sporophyte of a liverwort *(Marchantia)*. The enlarged elater shows spiral wall thickenings (modified after Parihar, 1961).

mechanisms and spore release occurs through irregular splitting of the capsule wall (Fig. 118). Most mosses have a special region, comprising the **peristome** and **operculum**, above the fertile section of the capsule (Fig. 119). This region is involved in regulating spore

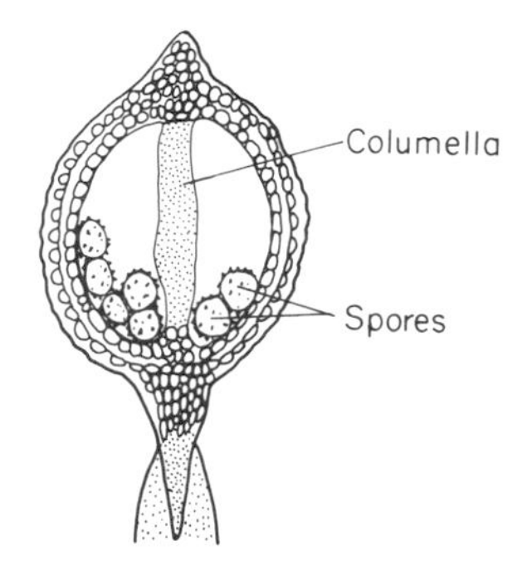

Figure 118. Longitudinal section through a simple moss sporophyte *(Ephemerum)*. Notice the sterile columella (modified after Parihar, 1961).

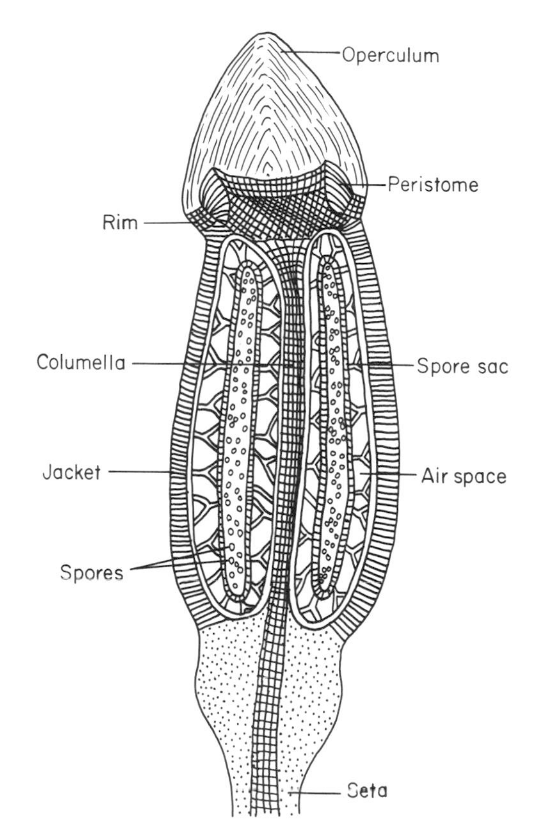

Figure 119. Longitudinal section through a moss *(Funaria)* capsule with an operculum and double peristome (modified after Parihar, 1961).

dispersal. In complex moss capsules (e.g., *Funaria*) there are two rows of peristome teeth beneath the operculum (Fig. 120). As the capsule dries the operculum is shed exposing the peristome. The teeth in each row are curved, narrow and triangular in shape. Exposed teeth are very sensitive to changes in atmospheric humidity. Each outer peristome tooth is a two layered structure with one outer layer and one inner layer. The outer layer lengthens when it absorbs moisture, and shortens as it dries. The inner layer is unaffected by drying or moistening. The net effect is that on absorbing moisture the outer peristome teeth bend over the capsule aperture and prevent spores from being shed. On drying the outer peristome teeth bend back and allow spores to fall freely into the atmosphere. The inner teeth act together as a sieve allowing spores to fall free gradually. The seta that supports the capsule in *Funaria* is also sensitive to humidity. As it dries the seta twists and swings the capsule around. In dry weather the whole mechanism works like a salt shaker. In wet weather spore liberation is prevented, presumably because wet spores disperse to a limited extent.

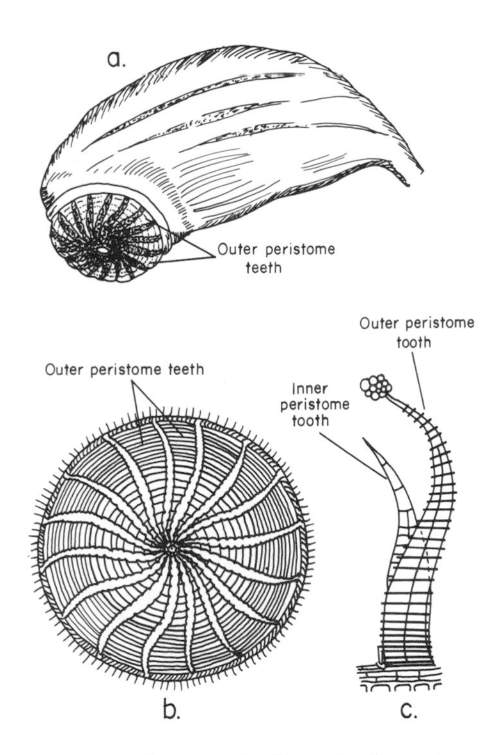

Figure 120. Peristome mechanism in *Funaria* A. mature capsule; B. outer ring of peristome teeth seen from above; C. inner and outer peristome teeth (modified after Parihar, 1961).

Some mosses have much simpler dispersal mechanisms than that of *Funaria*. In the peat moss, *Sphagnum*, an explosive system is employed. The capsules have a large central air space when mature. As the capsule dries the wall becomes impermeable to gasses and the thickening of the capsule jacket is such that the capsule can contract transversely, but not longitudinally. The result is a change of capsule shape from globose to cylindrical (Fig. 121) and a compression of trapped air to 4-6 atmospheres. The cells of the operculum do not shrink on drying and a strain is set up where it joins the capsule proper. Eventually the lid is blown away and the spores are shot into the air.

The capsule in hornworts (Fig. 122) may or may not contain a central column of sterile cells. True elaters with spiral thickenings are present in some species and these function in spore dispersal as they do in liverworts. Many hornworts have capsules with well developed stomata and internal air spaces. There is extensive development of photosynthetic tissue in these forms. However, water and minerals are always obtained from the gametophyte via the foot.

Spores and Protenemata

All bryophytes (except the moss *Macromitrium salakanum*) are homosporous. The spores have complex multilayered walls. The walls are cutinized and well adapted to terrestrial conditions. In the leafy

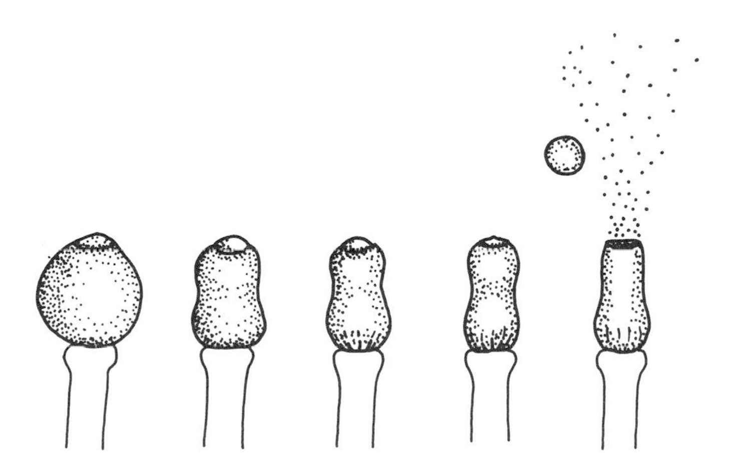

Figure 121. Spore dispersal by gas explosion in *Sphagnum* (moss) (modified after Ingold, 1965).

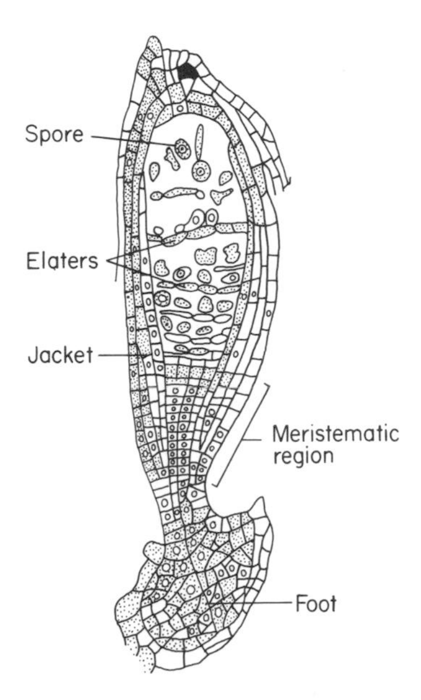

Figure 122. Mature sporophyte (longitudinal section) of a hornwort *(Notothylas)* showing spores, elaters and meristematic transition region (modified after Parihar, 1961).

bryophytes the spore germinates into a preliminary stage, the **protonema**, from which the adult gametophyte develops at a later stage. In the higher mosses the protonema is usually filamentous. The mature gametophyte develops from globose buds of 20-30 cells. Liverworts have protonemata that may be filamentous or parenchymatous (flat sheets or globose). Hornworts do not have protonemata.

Systematic Diversity of Bryophyta

The division Bryophyta is divided into three classes:

(a) Class Hepaticopsida (liverworts),

(b) Class Anthocerotopsida (hornworts) and

(c) Class Bryopsida (mosses).

CLASS HEPATICOPSIDA

The gametophyte is either a flat sheet of parenchyma or a leafy axis. Leafy forms usually have leaves (without midribs) arranged on the axis in two or three rows. The sporophyte has little or no photosynthetic tissue and lacks stomata. The capsule lacks a central column of sterile tissue.

CLASS ANTHOCEROTOPSIDA

This is the smallest group of bryophytes. The gametophytes are never leafy and consist of flat lobed thalli with no internal tissue differentiation. The sporophyte has no seta and meristematic activity is restricted to the transition region between the capsule and the foot. The capsule is always long and cylindrical. The capsule is equipped with stomata, internal air spaces and photosynthetic tissues. There is usually a central column of sterile cells in the capsule.

CLASS BRYOPSIDA

This class is the largest in the division. The gametophyte of mosses is differentiated into a protonema and an erect leafy shoot. The leafy shoot bears spirally arranged leaves and the gametangia. The sporophyte has a highly differentiated capsule which may, or may not, have functional stomata. Elaters are absent from moss capsules, but are common in the other two classes of bryophytes.

The evolutionary relationships of the bryophytes are unknown. The group shares many cytological features with green algae (Chlorophyta) and vascular plants (Tracheophyta). Bryophytes have much in common with other land plants: (a) functional stomata, (b) a water transport system, and (c) archegonia and antheridia with sterile jackets. These may point to common ancestry, or to convergent evolution to meet common needs.

12
Fungi

Fungi are never photosynthetic. This characteristic sets them apart from the other groups of organisms considered in this book. The inclusion of a chapter on non-photosynthetic organisms may seem anomolous in a discussion of plant diversity. However, the ecological relationships between plants and fungi are so close, and so important to man, that an introduction to fungal diversity is seen as an important component of this book.

The importance of fungi to human welfare is not immediately obvious. It has already been pointed out that the photosynthesis of plants on land can be severely limited by CO_2 concentrations in the atmosphere. Carbon dioxide that is photosynthetically fixed becomes incorporated into the organic components of plant cells. Forty to sixty percent of plant production ends up as cellulose. The great importance of fungi lies in their ability to digest cellulose as a heterotrophic carbon source. During the digestion of plant remains fungi release CO_2 back into the atmosphere. It is estimated that 85 billion tons of carbon as CO_2 are returned annually in this manner. Without fungal digestion of plant remains photosynthetic production would come to a stop within 20 years. The effect on human welfare is self evident.

Most of the carbon obtained by fungi for heterotrophic assimilation comes from plants. The plants providing the carbon may be alive or dead. Fungal attacks on live plants are the major cause of disease in human crops. Again the importance of this to mankind is obvious. Many of the major famines in human history resulted from fungal diseases of live crops.

Distinguishing Characteristics

The true fungi belong to the Kingdom Eukaryota. All fungi use organic sources of carbon and this characteristic separates the groups from bryophytes, vascular plants and seaweeds. Some phytoplankters are heterotrophic, but the vast majority are autotrophic. Most fungi are filamentous with extracellular walls. Cellulose is a rare component of fungal walls. Cellulose walls are obviously inappropriate in organisms that have extracellular cellulose digesting enzymes. Rigid cell walls preclude phagotrophy (engulfing food particles). Hence fungal heterotrophy involves the secretion of extracellular enzymes, absorption of the partially digested liquid substrate and intracellular assimilation.

This diagnosis of the fungi excludes a group of organisms known as slime molds or Myxomycetes. The slime molds resemble fungi in their heterotrophic nutrition. However, slime molds are phagotrophic and lack cell walls during their vegetative existence.

Vegetative Organization and Adaptation

Although most fungi are terrestrial, their vegetative organization is not clearly related to the problems that affect vascular plants and bryophytes. The non-reproductive structures of fungi generally have little resistance to desiccation. Adaptations for mechanical support are rare. Because of their nutritional mode, the environment surrounding feeding fungal cells is their food supply. The vegetative cells of most fungi live in their food and not in the atmosphere. Fungal cells are protected from desiccation by the tissues of other organisms.

The vegetative organization of fungi is closely related to nutritional life styles. Fungi are either **saprophytes** (living on dead organic matter), **parasites**, or they have mutual relations with other organisms. In the majority of saprophytes and parasites the basic vegetative unit of construction is the filament. Fungal filaments are called **hyphae** and a hyphal colony is called a **mycelium**. The development of a mycelium in laboratory plate culture is shown in Fig. 123. The feeding hyphae

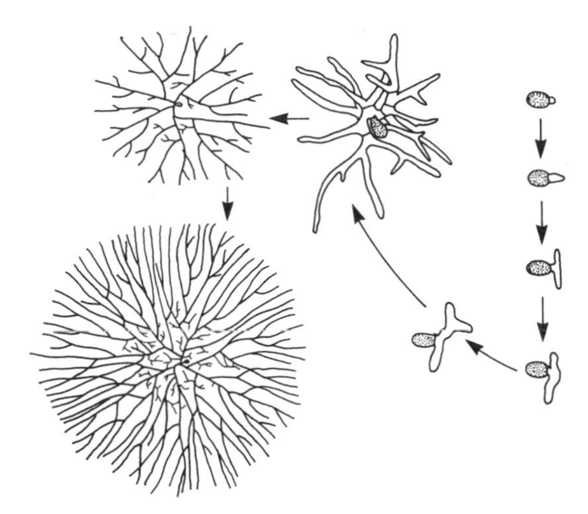

Figure 123. Development of a fungal hyphal mycelium from a spore. The two final stages of development are shown at a lower magnification than the first six stages (modified after Burnett, 1976).

are peripheral. The central mat exhausts its immediate nutrient supply. Growth occurs apically in the peripheral hyphae, while the central hyphae become firmly anchored in the substrate.

A diagram of the tip of a growing hypha is shown in Fig. 124. The major wall component is almost always **chitin**. Chitin is a polysaccharide consisting of N-acetyl-D-glucosamine residues. Apart from fungal walls, chitin is commonly found in arthropod exoskeletons. The outstanding characteristic of fungal walls is their strength and rigidity. Growth takes place only in a plastic apical hyphal zone. As hyphal tips grow they push forward with a tremendous mechanical pressure. There is an internal hydrostatic pressure which results from the high osmotic potentials of the cell contents. In addition, there is mechanical pressure which arises as a consequence of the synthesis of new cytoplasm. This pressure must be directed toward the apices since mature hyphal walls are rigid. The combination of an anchoring mycelium, rigid cell walls and apical growth pressure contributes to the ability of fungi to penetrate the hardest woods. The final factor of importance is the secretion of extracellular digestive enzymes.

Extracellular digestion in itself is inadequate for the complete breakdown of wood. Mechanical penetration is essential. In this respect fungi may be contrasted with bacteria. Saprophytic and parasitic bacteria also secrete digestive enzymes, but without mechanical penetration abilities, they are restricted to surface feeding only. This explains why the great bulk of dead plant tissue is recycled

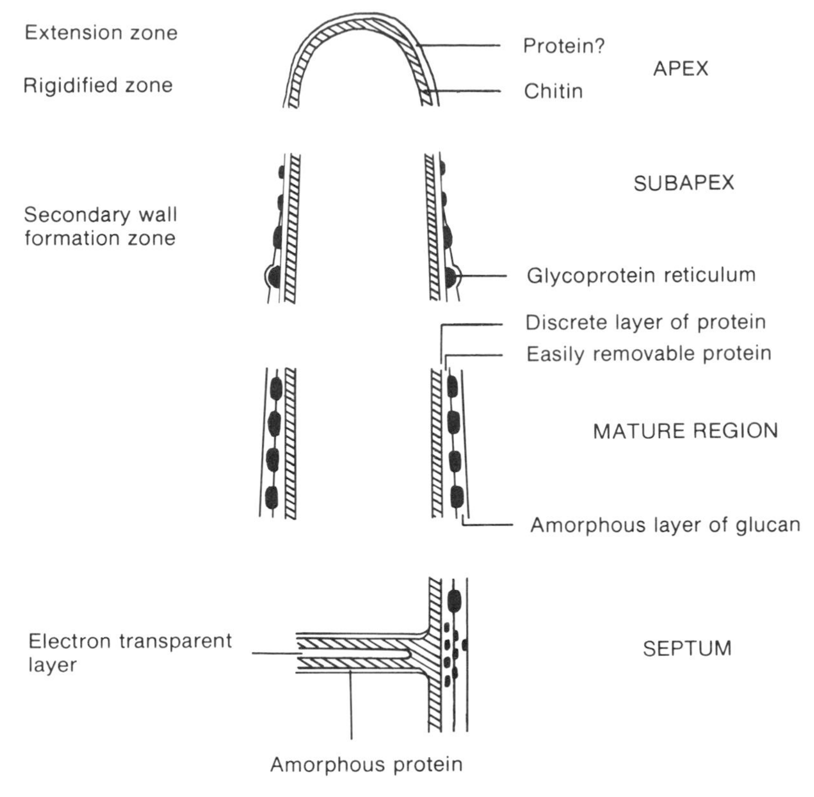

Figure 124. Apical organization in a septate fungal hypha (modified after Cole, 1981).

by fungi rather than bacteria.

The simple hyphal organization of saprophytic and parasitic fungi is directly related to feeding life styles. A filamentous construction is ideal for ramifying through live and dead food tissues. There are very few vegetative elaborations associated with these life styles. Among parasites special feeding hyphae called **haustoria** may develop (Fig. 125). A needle-like infection hypha grows into the host cell and develops into the haustorium. The haustorium is enveloped by the cell membrane of the host cell and does not make direct cytoplasmic contact.

Some fungi capture small worms in specially constructed traps (Fig. 126). If a worm noses into the trap the hyphae are stimulated and inflate in one-tenth of a second. The trapped animal dies and fungal feeding hyphae penetrate its body.

Differentiated vegetative multicellular structures are uncommon among parasites and saprophytes (but differentiated reproductive structures are common). Development of multicellular aggregations

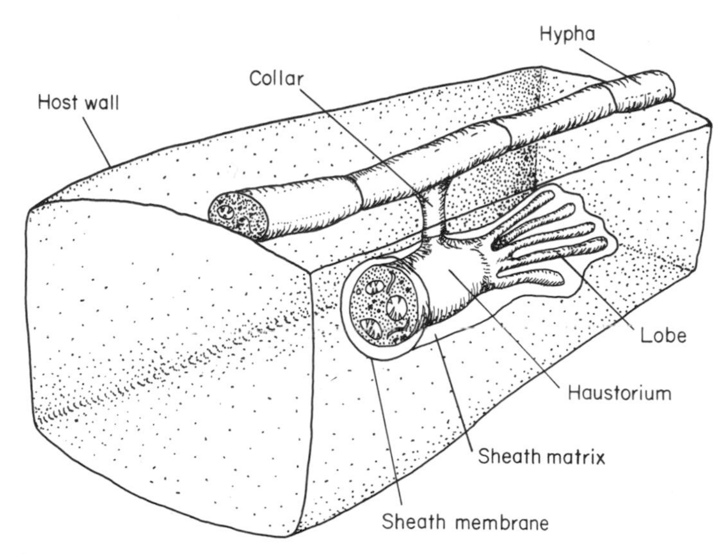

Figure 125. Specialized feeding haustorium in a parasitic fungus (modified after Neushul, 1974. © John Wiley & Sons, Inc.).

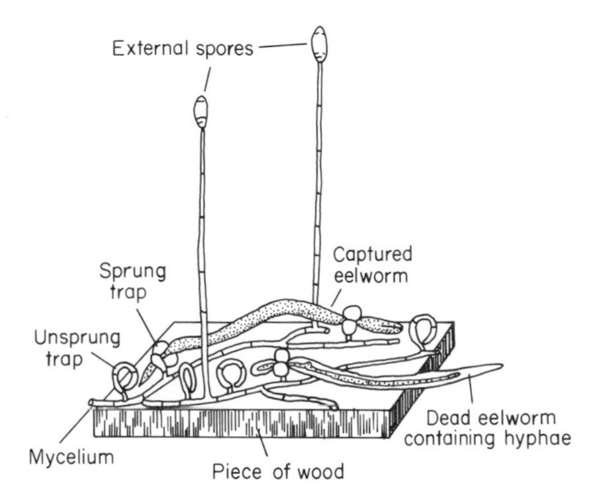

Figure 126. Eelworm trap mechanism in a hyphal fungus. Notice traps, feeding mycelia and sporangial apparatus (modified after Ingold, 1961).

is often associated with adverse conditions. For example, scarcity of nutrients encourages the formation of **mycelial strands** in the dry rot fungus *Serpula lacrymans* (Fig. 127). The strands are compose of coalesced hyphae. Within each strand several types of hyphae differentiate. **Vascular hyphae** have enlarged cell cavities, walls with annular or spiral thickenings and cross walls broken down so that the cells

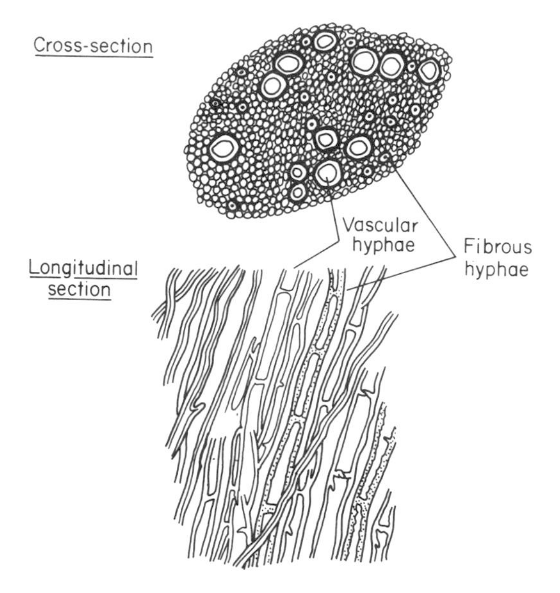

Figure 127. Hyphal differentiation in the mycelial strand of a dry rot fungus (modified after Burnett, 1976).

form open tubes. Mycelial strands of *S. lacrymans* definitely have a translocatory function, and the vascular hyphae may perform this role. The **fibrous hyphae** are narrow, thick walled and have small lumens. They appear to have a mechanical support function.

The **sclerotium** is another multicellular aggregation of vegetative hyphae (Fig. 128). The sclerotium is a resting structure that is highly resistant to water loss during periods of adverse conditions. The thick-walled outer rind is an impediment to drying out.

The most complex vegetative organization in fungi is found in species that have mutualistic life styles. **Lichens** are the prime examples. The lichen thallus is a composite structure having a fungal and an algal component (Fig. 129). The fungal component makes up the major part of the structure, but lichens bear little external resemblance to non-lichenized fungi. The fungal mycelium of lichens is strongly differentiated into pseudoparenchyma tissue layers (Fig. 129). The layer of algal cells is located within an upper cortical region of the thallus. The central medullary tissue layer consists of a loose web of hyphae. The medulla has considerable water holding capacity.

The fungal component of lichens has a close nutritional relationship with the photosynthetic algae that they surround. The hyphae

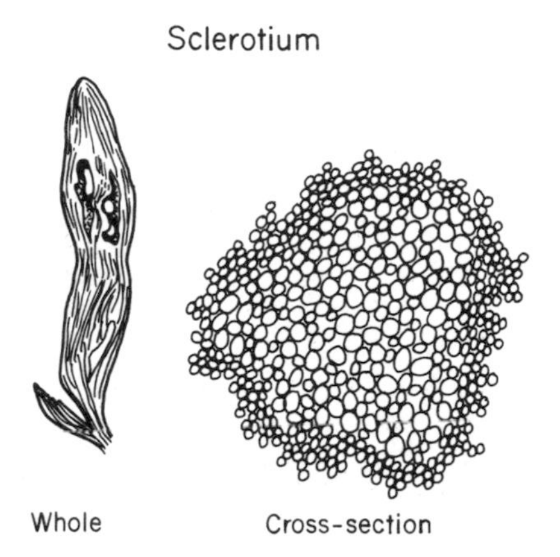

Figure 128. Multicellular sclerotium (thick-walled resting structure) (modified after Alexopoulos, 1962. *Introductory Mycology* © John Wiley & Sons, Inc. Reprinted by permission).

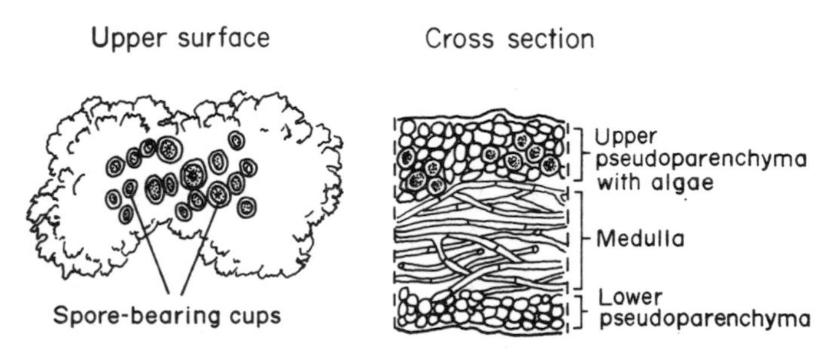

Figure 129. Surface and cross sectional views of an orange lichen. The algal component is lodged in the upper fungal pseudoparenchyma (modified after Ingold, 1961).

produce haustoria within the algal cells. There is no doubt that organic carbon is transfered from algae to fungal hyphae.

The thalli of lichens often have a complex morphology. Fructicose lichens are the most elaborate (Fig. 130). Lichens are not protected from desiccating terrestrial conditions within an organic food supply. Furthermore, lichens are often found under conditions of severe drought. There are, however, no special drought avoidance mechanisms. Lichens are drought tolerators and their water status depends entirely on atmospheric conditions. There are no special organs for water absorption or transpiration. When the thallus is wetted by rain, absorption occurs quickly. Much of the absorbed water is held within the cell walls and within the intercellular spaces

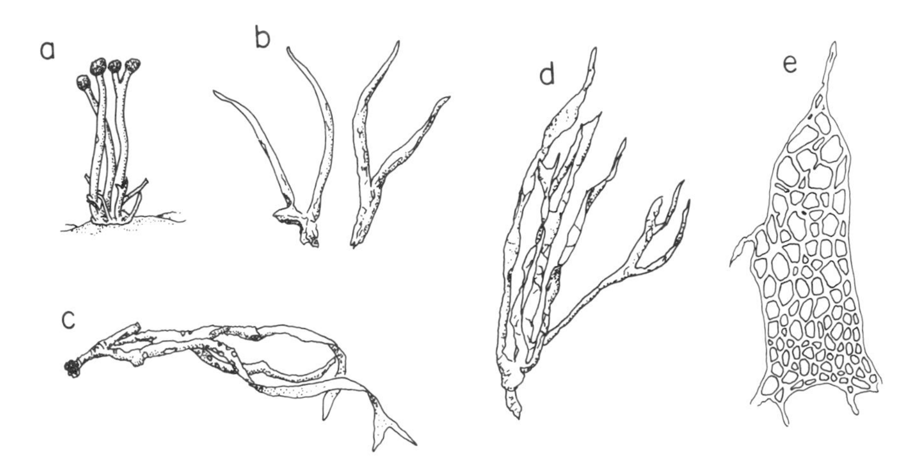

Figure 130. External morphologies of several fructicose lichens, a. *Pilophoro, cereolus,* b. *Thamniola vermicularis;* c. *Roccella fuciformis;* d. *Ramalina homalea;* e. *R. reticulata* (modified after Hale, 1967).

Reproductive Organization

There is an enormous diversity of reproductive mechanisms in fungi and only the briefest of introductions can be given here. The primary method of propagation is through spore production. Sexual reproduction also occurs. Both sexual and asexual modes of reproduction, and the life histories of fungi are related to habitat (aquatic or terrestrial) and to life style (saprophytic, parasitic or mutualistic).

ASEXUAL REPRODUCTION

In many aquatic fungi the spores are flagellate and are produced in specialized sporangia (Fig. 131). Some of the simpler terrestrial fungi also have flagellate spores. For example, *Synchitrium endobioticum*, which causes wart disease of potatoes, has flagellate spores that swim through the soil water. The great majority of terrestrial fungi have non-motile spores that are resistant to desiccating conditions. Nearly all spores, motile and non-motile, are agents of dispersal. A very few spore types act as resting stages that tide a species through periods of adverse environmental conditions.

The flagellate spores of aquatic fungi do act in dispersal, but swimming velocity is very low (0.25-1.0 cm/min) and most dispersal is via water currents. The importance of flagellar locomotion in spores lies in their ability to select a suitable substrate for settlement. The facility for substrate selection has a basis in the chemotactic

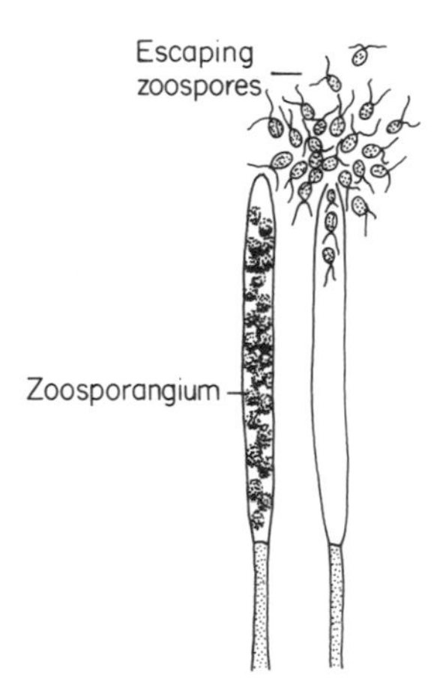

Figure 131. Sporangium and flagellate spores of a water mold (modified after Ingold, 1961).

behavior of spores. Motile spores of the water mold *Saprolegnia* actively swim toward a natural "bait" such as exposed muscle fibers in the leg of a dead fly. The flagellate spores of terrestrial parasitic fungi are also chemotactic. The spores of *Phytophthora parasitica* swim through soil water and are attracted to plant roots, especially those with small wounds.

Some terrestrial fungi (e.g., *Phytophthora infestans*) produce aerial sporangia. In wet conditions the sporangia produce biflagellate spores, but under dry conditions the whole sporangium detaches and germinates when in contact with a suitable host. This facultative production of various spore types affords considerable flexibility in mechanisms for dispersal. There are many species of aquatic fungi that have non-motile spores. For the most part these are considered to have had terrestrial ancestors. Among terrestrial fungi non-motile spores are dispersed by a huge variety of mechanisms. Generally we can distinguish spore dispersal by **carpophores** from other mechanisms. Carpophores are the fruiting bodies produced by the majority of terrestrial fungi (rust and smut parasites are notable exceptions). The mushroom is probably the best known carpophore, but most people are also familiar with bracket fungi and various other fruiting structures.

Carpophores consist of an **hymenium** which is made up of spore producing cells and other non-reproductive cell types. The hymenium is usually enclosed, supported or held aloft by sterile tissue. In a mushroom, for example, the hymenium is restricted to the gills on the underside of the cap. Most of the mushroom consists of highly organized sterile hyphae. The arrangement of sterile hyphae in carpophores of higher fungi represents the highest degree of anatomical differentiation found in the group. There are two basic types of carpophores, **basidiocarps** and **ascocarps**. Ascocarps are restricted to the largest group of terrestrial fungi, the **Ascomycota**. Basidiocarps are restricted to the **Basidiomycota**. Mushrooms are basidiocarps.

The greatest complexity and diversity among fungal carpophores is to be found in the Basidiomycota. Even the most complex types are simply coalesced aggregates of hyphae. The hyphae making up the basidiocarp have direct access to food supplies. The carphophores are often connected to a feeding mycelium by mycelial strands or other differentiated structures. The relationships among the feeding mycelium, carpophore and the dispersing spore cloud are shown in Fig. 132. The sectional view of the mushroom shows the gills. A high magnification examination of a gill section reveals the structure of the hymenium (Fig. 133). The spore-producing cells (**basidia**) each

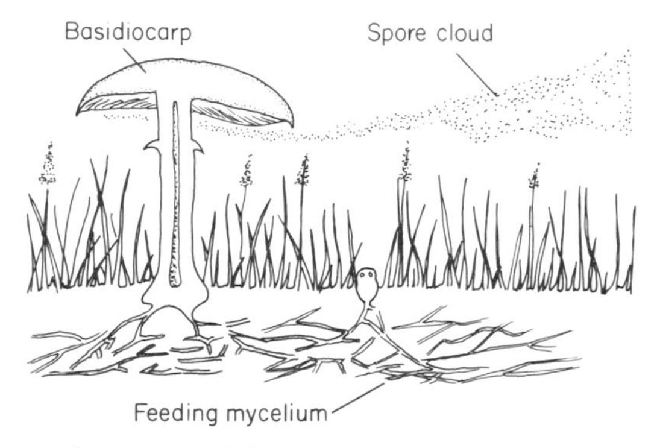

Figure 132. Basidiocarp and feeding mycelium of a mushroom (modified after Ingold, 1961).

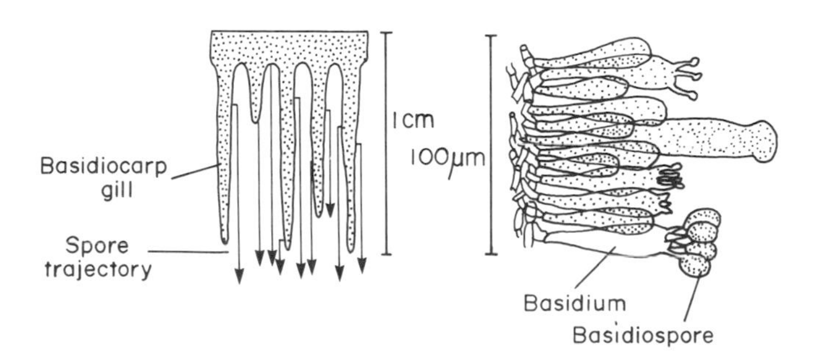

Figure 133. Vertical sections of basidiocarp gill showing spore trajectories and details of hymenium organization (modified after Ingold, 1971 by permission of Oxford University Press).

form four **basidiospores**. The spores are shot away from the basidium to a distance that is normally 0.10-0.15 mm. The spores then fall freely away between the gills (Fig. 134). The exact mechanism by which basidiospores are discharged is unclear. In order to allow free fall of spores the gills must be exactly vertical. The mushroom stalk responds to gravity and light so that the cap is positioned correctly for spore fall. In addition, most mushrooms have gills that can move in response to gravity to finely adjust their orientiation.

Within basidiocarps three basic hyphal types may be recognized (Fig. 135):

(a) **generative hyphae**—these are thin-walled, branched and often inflated; they give rise to other hyphal types,

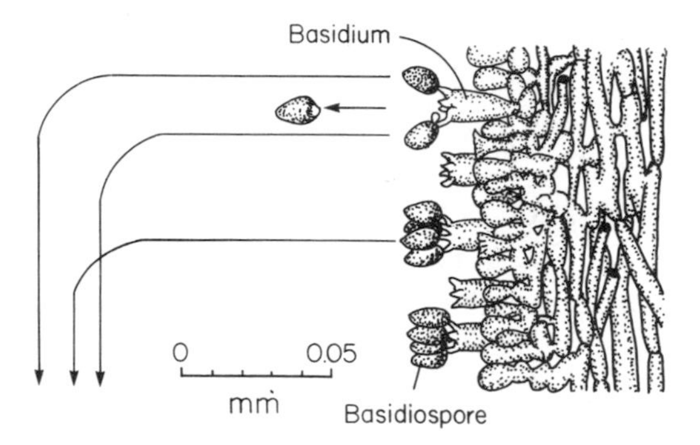

Figure 134. Violent dispersal and free-fall trajectories of basidiospores (modified after Ingold, 1971).

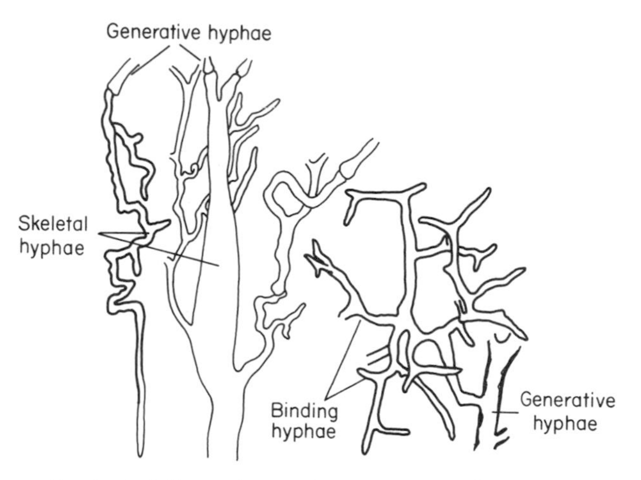

Figure 135. Hyphal differentiation found in basidiocarps (modified after Burnett, 1976).

(b) **skeletal hyphae**—these are thick-walled and usually unbranched; they give a rigid framework that is locked together by the binding hyphae,

(c) **binding hyphae**—these are narrow, thick-walled and much branched.

Translocating hyphae have been recognized in the stalks of some basidiocarps. This represents a further level of differentiation. Finally, of course, there is differentiation of the hymenium.

There are three kinds of ascocarps in the Ascomycota (Fig. 136). In the **apothecial** ascocarp the hymenium is exposed on an open surface. The sporangia that produce the non-motile spores are called **asci**. In the **perithecial** construction the asci are enclosed within a flask-shaped structure. Each ascus discharges its spores in turn through the narrow neck of the perithecium. **Cleistothecia** are very small ascocarps containing one or a few asci. Cleistothecia represent resting stages in the life histories of some ascomycetes.

In many ascomycetes the ascus is an explosive spore dispersal device (Fig. 137). A high internal pressure develops as the osmotic concentration of the cells increases. The ascus eventually bursts open apically. The point at which an ascus opens may be a hinged lid, an apical slit, a detachable cap or a minute pore.

Ascospores are usually discharged to a distance of 0.5-2.0 cm into the turbulent flow of air above the boundary layer. In this manner the spores of Ascomycetes are dispersed by air currents.

In many terrestrial fungi the sporangia are not produced within the tissues of a differentiated carpophore. Spores are often formed on erect hyphae that grow up from the feeding mycelium into the turbulent air currents (Fig. 138). Spores are shed and dispersed passively from such structures. In addition, active water squirting de-

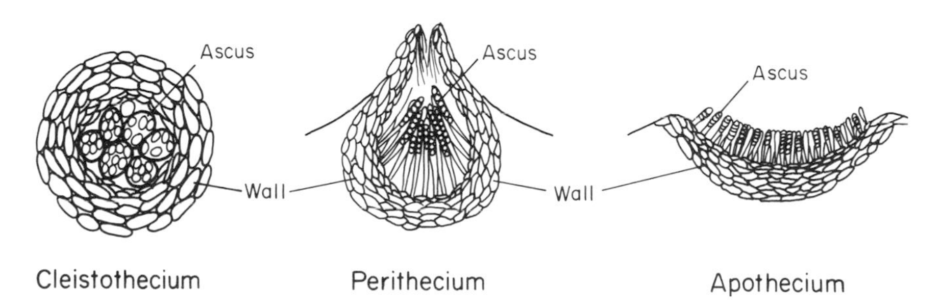

Figure 136. Three types of ascocarps found in the Ascomycota (modified by permission from Pritchard, and Bradt, 1984. *Biology of Nonvascular Plants.* The C. V. Mosby Co.).

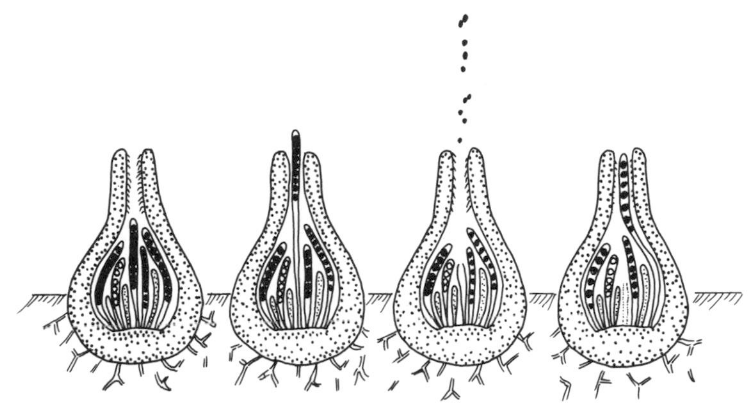

Figure 137. Explosive spore discharge from a perithecial ascocarp (modified after Ingold, 1971 by permission of Oxford University Press).

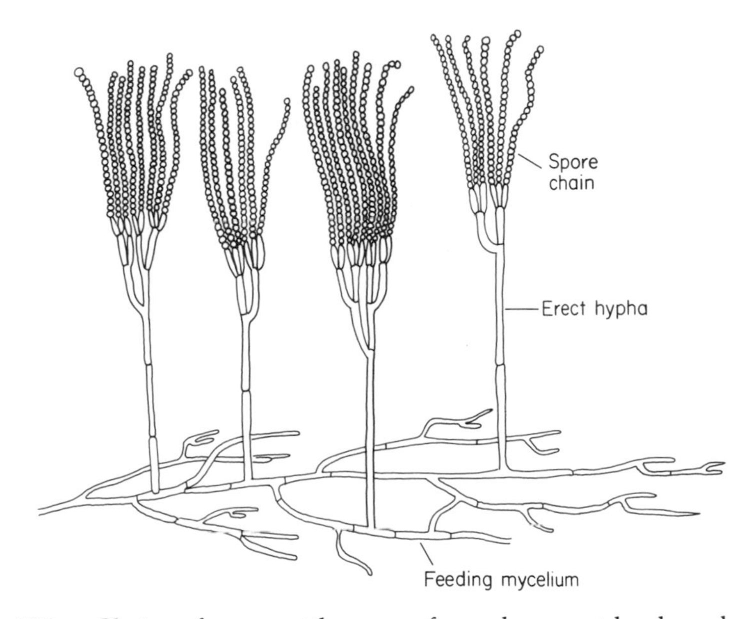

Figure 138. Chains of non-motile spores formed on erect hyphae above a feeding mycelium (modified after Ingold, 1961).

vices are employed in some species that lack carpophores (Fig. 139).

Many lichens have an ascomycete fungal component. Ascocarps and ascospores are produced in these forms. In order to form a new lichen thallus, a settled ascospore must make contact with an algal symbiont. To ensure synchronized reproduction of fungal and algal components, many lichens produce propagules containing both symbionts.

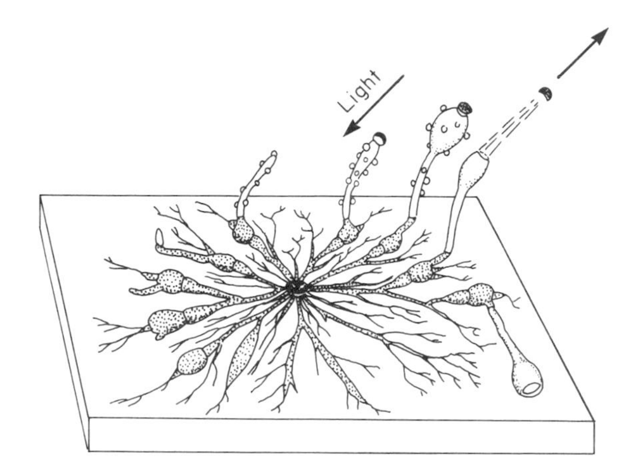

Figure 139. Water-squirting spore dispersal device of *Pilobolus* (modified after Ingold, 1971).

SEXUAL REPRODUCTION

There is an enormous diversity of sexual processes in the fungi. The basic components are nuclear fusion and meiosis, as in other groups of eukaryotes. Diversity arises in the ways in which nuclei are brought together for fusion. Two aspects of this diversity are considered here, genetic mating systems and morphological diversity among gametes and gametangia.

The genetic mating system found in many fungi is called **homomictic**. In homomictic forms self-fertilization is possible and occurs regularly. In fungi that are **dimictic** there are two genetic mating strains (usually called + and −). Fertilization can occur only between nuclei of two different mating strains. Generally in dimixis mating strain type (+ or −) is determined by a single genetic locus with two alleles.

In some of the Basidiomycota sexual differentiation may involve hundreds of genetically different strains involving many loci. The system works as follows. If there are two mating loci, *A* and *B*, each with four alleles (denoted by superscripts 1-4), sexual fusion can only occur when all the alleles are different in the two mating partners. For example, mating may occur if one of the mating partners is A^1B^2 and the other is A^2B^3. Mating does not occur when the genotypes are A^1B^2, A^1B^3, or in any combination where there are shared alleles. This is a relatively simple example. There may be more than three loci involved, each with tens or hundreds of alleles. Provided the mating alleles differ in two mycelia, sexual fusion is possible. A system such as this clearly favors outbreeding. Where there are many genetically determined mating types the system is called **diaphoromixis**. Equally complicated genetic breeding systems occur in some flowering plants (e.g., primroses).

In the Ascomycota and Basidiomycota transfer of nuclei from one mating strain to another is not immediately followed by nuclear fusion. The gamete nuclei are, of course, haploid, but because nuclear fusion and cytological fusion are not synchronized, the product of sexual transfer is a cell with two haploid nuclei. This cell generates a mycelium with haploid nuclei of two genotypes called a **dikaryon**. In other eukaryotes cell fusion is almost always followed immediately by nuclear fusion. In the Ascomycota and Basidiomycota nuclear fusion is delayed until just before spore formation takes place (usually in a carpophore). Nuclear fusion is followed immediately by meiosis in these cases, so that the spores are haploid and so are the mycelia that they give rise to. In most fungal groups there is little evidence for the persistence of a diploid phase. Most mycelia are haploid, but because of the breeding systems, there may be several haploid genotypes among the nuclei of a single organism.

Sexual fusion among compatible mating strains is not the only mechanism by which genetic diversity arises in fungi. There exists among many of the terrestrial fungi a condition known as **heterokaryosis**. A heterokaryon is a mycelium in which genetically different nuclei co-exist. Unlike dikaryons, heterokaryotic mycelia may contain three or more nuclear types, and these can be diploid or haploid. Heterokaryosis arises in one of three ways. A mutation may occur within an originally genetically homogeneous mycelium and the mutant and non-mutant nuclei co-exist. Alternatively, genetically distinct mycelia may fuse leading to the co-occurrence of two or more nuclear types in a single mycelium. Diploid nuclei can arise by **diploidization** or somatic fusion of two nuclei within the mycelium. The importance of heterokaryosis lies in the genetic diversity that is maintained and propagated through the formation of genetically diverse spore types.

The process of diploidization may sometimes be followed by the breakdown of diploid nuclei to the haploid condition. The process is poorly understood, but is apparently different from classic meiotic chromosomal segregation. The term **parasexuality** is applied to the sequence of heterokaryosis, fusion of genetically dissimilar nuclei and recombination and segregation of the diploid nuclei.

In all sexually reproducing organisms nuclear fusion is preceded by cytoplasmic fusion. Thus fertilization has two components. The structures involved in cytoplasmic fusion vary greatly among different fungal groups. Five general mechanisms by which fungal nuclei are brought together (for fusion) may be recognized:

(a) **planogametic fusion**—in the simplest condition morphologically identical flagellate gametes fuse (isogamy); anisogamy and oogamy occur in forms with motile gametes,

(b) **gametangial contact**—in this mechanism the gametes are never released and gamete nuclei are transferred between gametangia (Fig. 140); gametangia may grow toward one another using gaseous hormones as direction signals; the gametangia do not fuse or lose their identity,

(c) **gametangial fusion**—in this mechanism there is complete fusion of contacting gametangia (Fig. 141),

(d) **spermatization**—minute non-motile male gametes called **spermatia** are carried passively to receptive female structures to which they become attached; the gamete nucleus of the spermatium is injected into the female structure,

(e) **somatogamy**—no gametangia or gametes are produced in many Basidiomycota. Fusion and nuclear transfer occur in ordinary undifferentiated hyphae of suitable mating strains.

Obviously planogametic fusion mechanisms are adaptive in aquatic fungi. Similarly, the absence of differentiated motile gametes can be seen as adaptive in terrestrial forms. In some species the method of sexual fusion is not clearly related to aquatic or terrestrial life styles.

There are four major fungal life histories involving meiosis and cytoplasmic/nuclear fusion (Fig. 142). The **diploid** cycle is quite rare and occurs only in some simple forms. In *Phytophthora*, for example, the hyphal mycelium is diploid, and haploid nuclei are thought to occur only in the gametangia. The **haploid** cycle is very common and, in fungi with this type of life history, diploid nuclei are found only in the zygotes. Meiosis quickly establishes a haploid condition that is found in the feeding mycelia. An alternation of **diploid** and **haploid** phases is found in some of the simpler fungi (e.g., *Allomyces*). The haploid hyphae produce gametes. Fusion results in a zygote which develops into diploid hyphae. These hyphae produce spores mitotically and meiotically. Mitotically produced spores re-establish the diploid phase. Meiotic spores produce a new haploid phase.

The **haploid/dikaryotic** cycle involves the production of two feeding mycelia. The haploid mycelium produces a dikaryotic mycelium through cytoplasmic fusion. Frequently the carpophores of

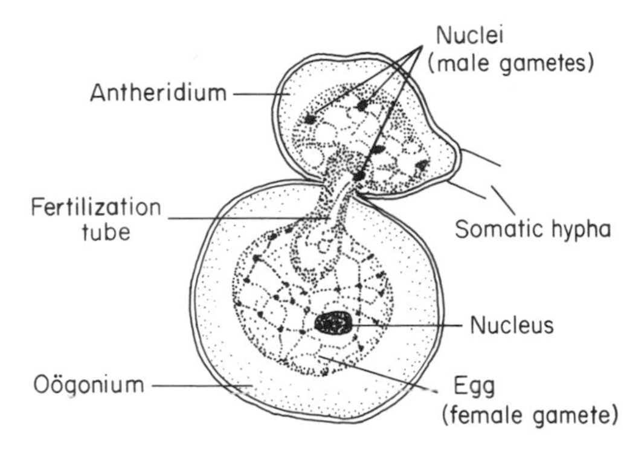

Figure 140. Sexual fusion through gametangial contact (modified after Alexopoulos, 1962. *Introductory Mycology.* © John Wiley & Sons, Inc. Reprinted by permission).

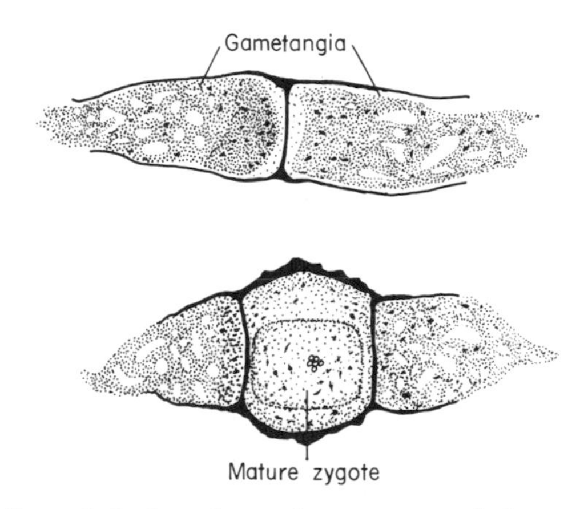

Figure 141. Sexual fusion through gametangial fusion (modified after Alexopoulos, 1962. *Introductory Mycology.* © John Wiley and Sons, Inc. Reprinted by permission).

higher fungi are produced by dikaryotic hyphae. Nuclear fusion in the hymenium is followed immediately by meiosis and the production of haploid spores. In this type of life history the diploid condition is restricted to the zygote nucleus and no separate diploid mycelium exists.

Systematic Diversity of Fungi

The members of the fungi are divided among the last six divisions listed in Table 2. The table shows only one of a large number of alternate classifications. The fungal divisions fall broadly into two

1. HAPLOID

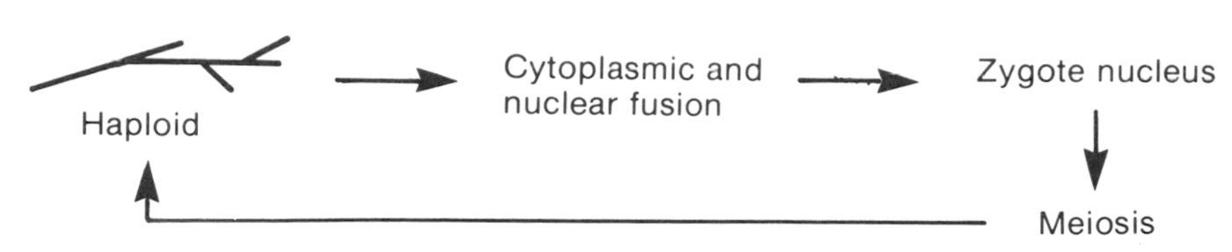

2. HAPLOID-DIKARYOTIC

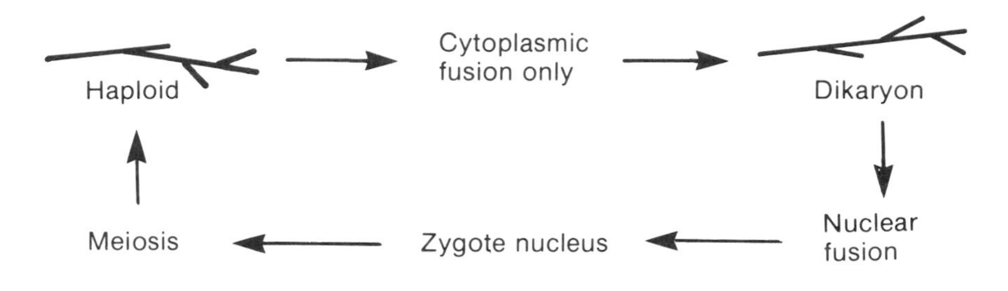

3. HAPLOID-DIPLOID

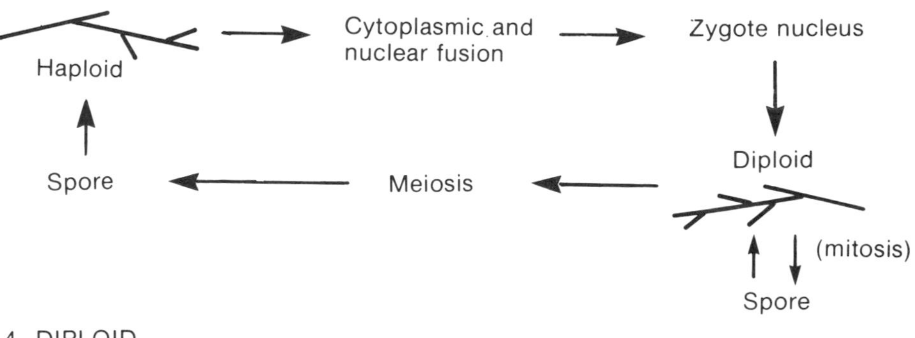

4. DIPLOID

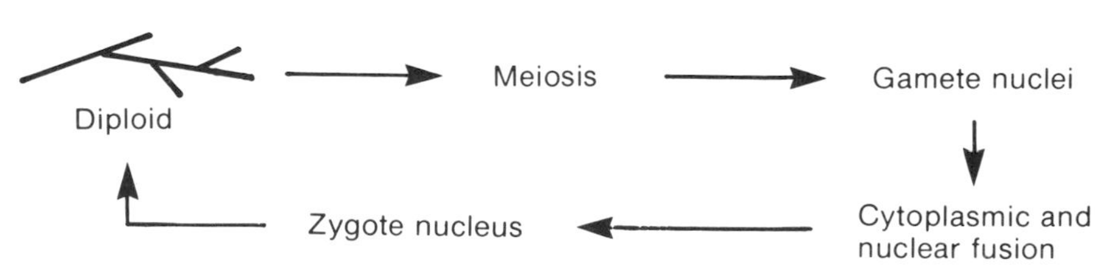

Figure 142. Four sexual life histories found in the fungi.

groups called lower fungi and higher fungi. The lower fungi are either unicellular, or in hyphal forms, coenocytic. The asexual spores of lower fungi are produced internally within sporangia. The higher fungi include those forms in the Ascomycota and Basidiomycota that produce differentiated carpophores. The spores of higher fungi are produced externally and all of them are non-motile. Unicellular higher fungi are rare (yeasts are a notable exception). Hyphal forms are septate, that is cross walls separate individual cells in the filaments.

DIVISION CHYTRIDIOMYCOTA

Members of this group are simple water molds that produce motile flagellate spores with a single flagellum. The major cell wall component is chitin.

Morphological diversity is very limited in the group. Multinucleate unicells called **chytrids** represent one morphological type (Fig. 143). The multinucleate sac bears feeding rhizoids that penetrate host cell walls. Chytrids are common parasites of diatoms.

A second morphological expression found in the division is the coenocytic filament. Fungi with this type of morphology are called water molds. These molds are usually saprophytic rather than parasitic.

Asexual reproduction in the Chytridiomycota is through flagellate spores. In unicellular forms the cell is converted into a sporangium. In mycelial forms a differentiated sporangium produces spores (Fig. 144). Life histories among the unicellular chytrids are haploid (Fig. 142). In the water molds haploid/diploid life histories occur (Fig. 142). In both life histories sexual fusion is generally isogamous and nuclear fusion follows quickly after cytoplasmic fusion.

DIVISION OOMYCOTA

Members of this group are water molds and downy mildews. All of them develop extensive coenocytic mycelia. The cell walls are

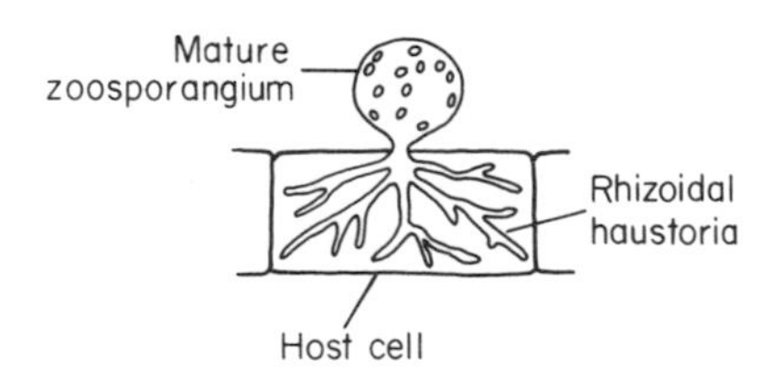

Figure 143. Feeding haustoria and zoosporangium of a parasitic chytrid (modified after Pritchard and Bradt, 1984).

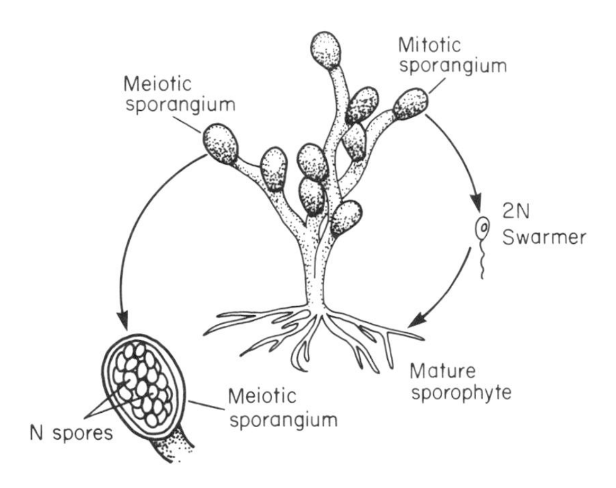

Figure 144. Sporangial differentiation in *Allomyces* (modified after Pritchard, and Bradt, 1984. *Biology of Nonvascular Plants.* The C. V. Mosby Co.).

usually cellulosic and rarely contain chitin. Flagellate spores lack cell walls and have two laterally inserted flagella. These characteristics differentiate the group from the chytridiomycetes.

The two main groups within the Oomycota (water molds and downy mildews) have rather different vegetative and reproductive characteristics reflecting their different life styles. Most of the molds are saprophytic, whereas the downy mildews are major terrestrial parasites. Well differentiated feeding haustoria are found in the downy mildews, but not in the water molds (even among parasitic species).

Asexual reproduction in aquatic forms typically involves the production of flagellate spores. The terrestrial mildews produce spores that are wind-dispersed and non-motile (Fig. 145). Spores that land in a drop of water on a potential host behave in one of two ways. The spore may germinate directly to produce an infectious mycelium. Alternatively, the cytoplasmic contents divide up to form flagellate spores. These motile bodies are released into the infection. After a short period of motility each encysts and then forms a short germ tube that produces a new mycelium. In some genera of downy mildews the spores never act as sporangia.

Sexual fusion in the Oomycota is by gametangial contact (Fig. 140). Even in aquatic forms free-living gametes are absent. Gamete nuclei appear to be produced meiotically and the zygote nuclei give rise to a diploid vegetative mycelium (see Fig. 142 Diploid cycle). In other groups of fungi the feeding mycelium is haploid or dikaryotic.

Both the water molds and the downy mildews are important

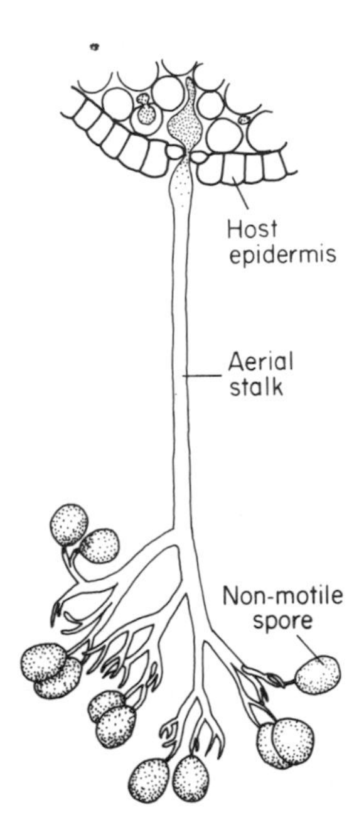

Figure 145. External non-motile spores produced by the parasitic fungus *Perenospora parastica* (modified after Ingold, 1961).

parasites. Some *Saprolegnia* (water mold) species are serious parasites of fish. The mildews of the Oomycota have been the causes of major famines in the past. The great Irish potato famine of the last century was caused by an oomycete. Another oomycete fungus is estimated to have destroyed 5% of Australia's commercial timber resource through root rot disease.

DIVISION ZYGOMYCOTA

Members of this group are terrestrial molds that do not have flagellate spores. Sexual reproduction is by gametangial fusion and results in the formation of a thick-walled **zygosporangium** (Fig. 141). The zygosporangium is a resistant resting stage.

The vegetative feeding mycelium of some of the bread molds in this division is differentiated into a submerged portion from which aerial hyphae called **stolons** also develop (Fig. 146). Where an aerial stolon makes contact with the substratum it becomes attached by a cluster of coarse rhizoids. Sporangial stalks grow up from the stolon above the point of rhizoidal contact. The submerged hyphal mycelium is a feeding structure whereas the clusters of rhizoids are for attachment only. The stolons have a vegetative propagation function similar to that of strawberry runners.

Asexual reproduction in the zygomycetes is by non-motile spores that are often produced in sporangia elevated above the substrate on sporangial stalks (Fig. 146). In many forms the spores are passively dispersed in the turbulent wind currents into which the sporangial stalks grow. In addition, water-squirting devices are to be found among genera that live on animal dung.

The zygomycetes are generally dimictic, and cytoplasmic fusion occurs between gametangia of different mating strains. The hyphae are haploid. Diploidy is established only in the zygote after nuclear fusion. Several nuclear fusions occur after gametangial fusion has occurred. The whole structure containing the zygote nuclei encysts forming a characteristic thick walled resting stage. Meiosis occurs in the encysted stage. After germination of the encysted stage, short sporangial stalks are produced and haploid spores are released.

The Zygomycetes live a variety of terrestrial life styles. Many are saprophytic molds, but others are parasites (notably of insects). Some members of the division have mutualistic relations with plant roots. The fungi actually penetrate root cells and have been shown to benefit the plant by enhancing nutrient uptake in mineral poor soils.

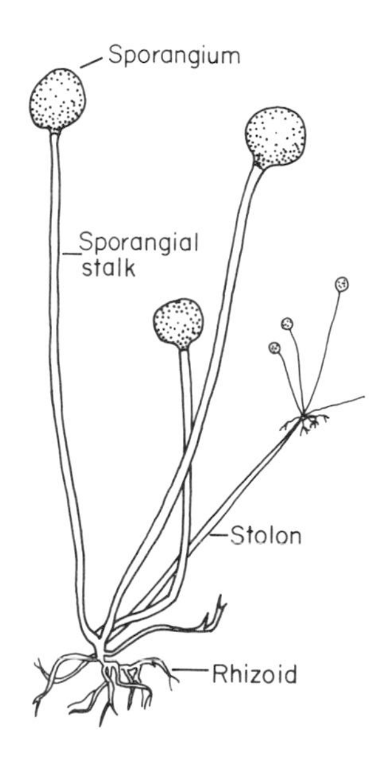

Figure 146. Morphological differentiation in *Rhizopus* (from *Nonvascular Plants: An Evolutionary Survey* by Scagel, Bandoni, Maze, Rouse, Schofield and Stein. © 1982 by Wadsworth Publishing Co. Reprinted by permission of the publisher).

DIVISION ASCOMYCOTA

This division is the largest fungal group. Apart from the great number of saprophytic and parasitic taxa, almost all of the 20,000 or more species of lichens have an ascomycete component. Together with the Basidiomycota and Deuteromycota, the ascomycetes form a loose group called the higher fungi. Most of the members of this group are terrestrial. The only structure which differentiates ascomycetes from all other fungi is the **ascus** (see Fig. 136). Ascus formation follows sexual cytoplasmic fusion (Fig. 147). After sexual contact and transfer of nuclei a dikaryotic condition is established. Hyphae grow out from the apex of the oogonium (called an **ascogonium** in this group), and each cell of each hypha contains two nuclei (they are dikaryons). Nuclear fusion occurs in the apical cells of the dikaryotic

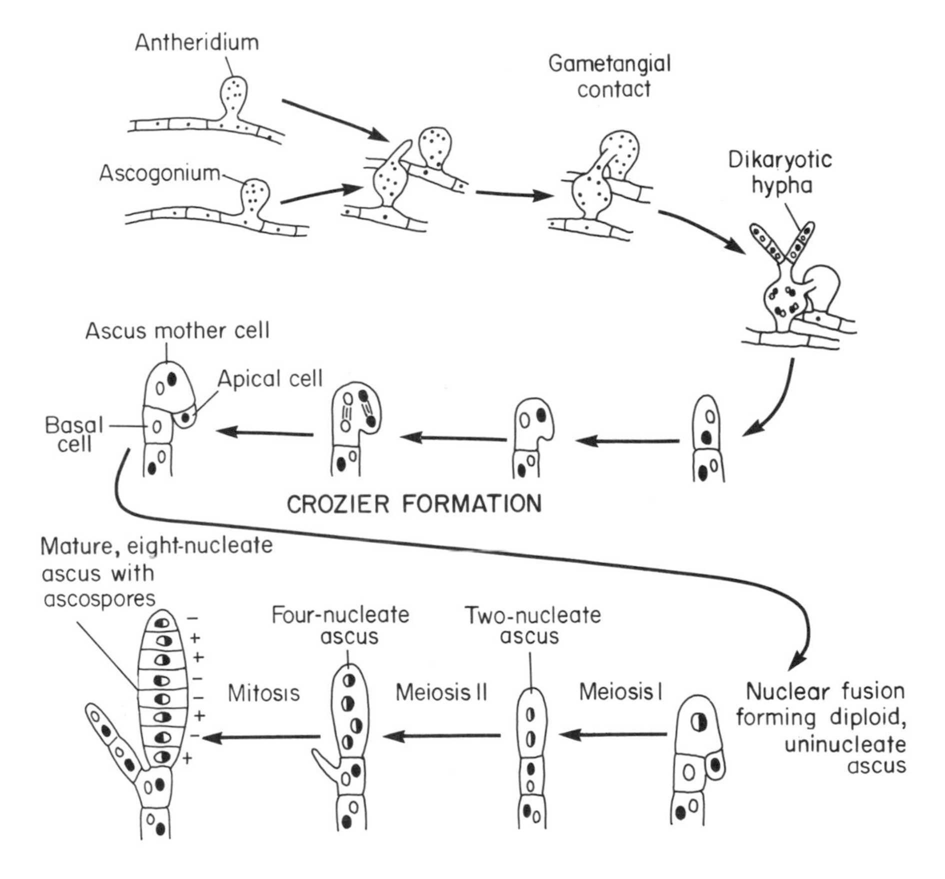

Figure 147. Development of haploid ascospores following cytoplasmic fusion, growth of dikaryotic hyphae, nuclear fusion, meiosis and mitosis (modified after Pritchard, and Bradt, 1984. *Biology of Nonvascular Plants.* The C. V. Mosby Co.).

hyphae. Meiosis occurs, immediately leading to the formation of four haploid nuclei. Mitotic division of each nucleus takes place so that eight haploid nuclei are contained within the single cell now known as an ascus. Eight **ascospores** differentiate, each with a haploid nucleus.

Ascomycetes are further characterized by a mycelium that is septate (divided into cells by cross walls). This trait is shared by the Basidiomycota and Deuteromycota, but the lower fungi are all coenocytic or unicellular. The cross walls of higher fungi are perforated at the center so there is cytoplasmic contact from cell to cell. The cross wall perforations allow nuclear migration, and this is of importance in maintaining a dikaryotic mycelium.

There are two morphological types of ascomycetes. The smaller group has unicellular members called yeasts. The larger group is mycelial. In the mycelial forms differentiation of various tissue types is common (e.g., mycelial strands and various other forms of pseudoparenchyma). The most complex tissue differentiation is to be found in carpophores called **ascocarps**. Not all ascomycetes form ascocarps.

Asexual reproduction is very varied in the Ascomycota. Simple fission and budding occur in the unicellular yeasts. In mycelial forms non-motile spores are formed externally (not in sporangia). These spores are generally borne on aerial hyphae that project up into the turbulent air currents. In some forms the spore-bearing hyphae are formed in differentiated fruiting bodies (Fig. 148). It must be emphasized that the discussion of asexual reproduction to this point has excluded the formation of ascospores, asci and ascocarps. These three structures develop as a consequence of sexual fusion.

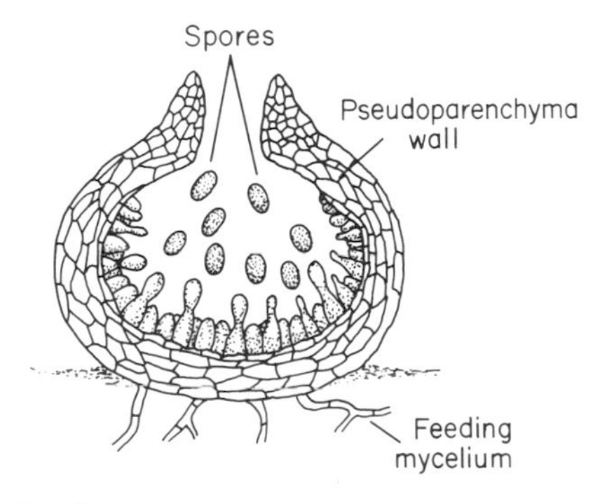

Figure 148. Longitudinal section of an ascomycete differentiated fruiting body showing formation of non-motile spores (from *Nonvascular Plants: An Evolutionary Survey* by Scagel, Bandoni, Maze, Rouse, Schofield and Stein. © 1982 by Wadsworth Publishing Co. Reprinted by permission of the publisher).

It is convenient to begin a consideration of the sexual cycle of Ascomycota with the germination of an ascospore. The product of the germination is a mycelium of uninucleate haploid cells. Sexual compatability is homomictic or dimictic in these haploid mycelia. Cytoplasmic fusion occurs in a variety of ways (gametangial copulation, gametangial contact, spermatization or somatogamy). The consequence of cytoplasmic fusion is a dikaryotic condition, followed by nuclear fusion, then meiosis and the development of ascospores. The life history is therefore haploid/dikaryotic.

The ascomycetes are of enormous ecological and economic significance. They have a major role in the recycling of key elements that are locked in the organic products of plant metabolism. Fungi in this group are able to decompose a large variety of organic polymers, including cellulose, petroleum products and even some plastics and polyesters. In their lichen guise ascomycetes are often pioneer occupants of barren substrata. Lichens contribute significantly to the build up of organic litter in which plant roots later develop. Lichens are a major vegetation component in huge areas of Arctic tundra.

The economic importance of ascomycetes is very great. The baking and brewing industries are based on the activities of yeasts. Ascomycetes are important sources of antibiotics. The parasitic species are serious pests of timber and crops. Dutch elm disease and chestnut blight are ascomycete diseases. Apple scab, powdery mildew and a wide variety of other crop diseases are caused by parasitic members of the division.

DIVISION BASIDIOMYCOTA

Members of this division are diagnosed by the production of **basidiospores** which are borne externally on a club-shaped hyphal tip called a **basidium** (Figs. 133, 134). Basidiospores are often thought of as homologs of ascospores since both types develop in a comparable manner. Most of the commonly known fungi belong to the Basidiomycota. Common examples are mushrooms, toadstools, puffballs, bracket fungi and the serious parasites known as rusts and smuts.

The vegetative mycelium of basidiomycetes consists of well-developed hyphae with cross walls. The cross walls of all members of the group (excluding rusts and smuts) is very distinctive (Fig. 149). The central pore is surrounded by a doughnut-shaped thickening. Each pore is covered by a septal pore cap composed of endoplasmic reticulum. This type of structure is especially characteristic of the dikaryotic phase. Dikaryotic hyphae with distinctive septal pores have a further distinguishing characteristic known as the **clamp**

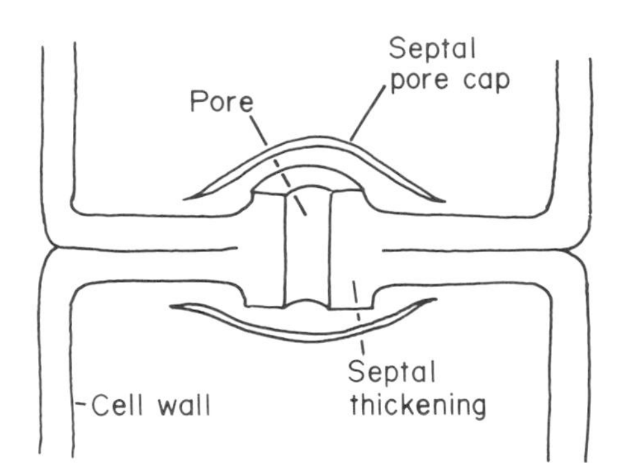

Figure 149. Elaborate cross wall perforations found in basidiomycetes (modified after Pritchard and Bradt, 1984. *The Biology of Nonvascular Plants.* The C. V. Mosby Co.).

connection (Fig. 150). Formation of clamp connections ensures the maintenance of the dikaryotic condition in the newly forming cells at the hyphal apex.

The haploid stages of basidiomycetes are generally hyphal. The dikaryons are all hyphal and in most species they are independent of the haploid mycelium. The independent, long-lived dikaryotic mycelium is a basidiomycete characteristic that differentiates most forms from the ascomycetes.

The vegetative mycelia of basidiomycetes may form complex tissues. Thick strands known as **rhizomorphs** (Fig. 151) are often differentiated. Growth is coordinated so that the whole unit grows from an apical meristem.

Asexual reproduction occurs in the basidiomycetes, but not as frequently as in other fungal groups. There are two main types of asexual reproduction, fragmentation of the hyphal mycelium and external production of non-motile spores. In the rusts and smuts spores are often produced in small fruiting bodies that have a similar organization to those in the Ascomycota.

In most basidiomycetes no sex organs are formed. Somatic fusion of haploid hyphae leads to the formation of a dikaryotic condition. In some cases detached somatic cells (produced by hyphal fragmentation) are involved in sexual fusion. The rusts do form specialized hyphae whose sole function is sexual reproduction by spermatization. Sexual fusion occurs between compatible hyphae. There are monomictic, dimictic and diaphoromictic species in the division.

The haploid sexual mycelium of basidiomycetes is called the primary mycelium. Sexual fusion establishes an independent dikaryotic

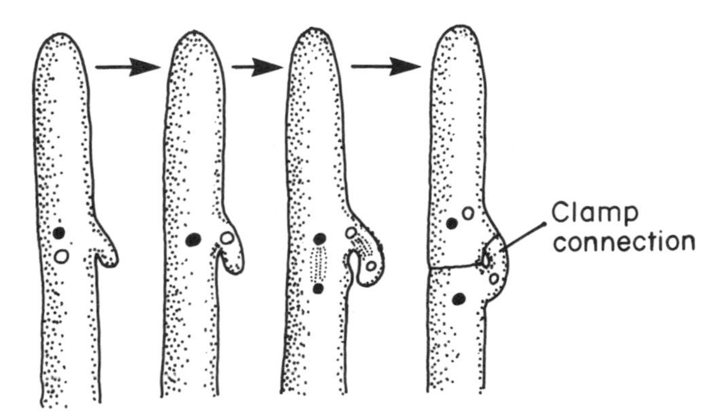

Figure 150. Apical cell division and clamp connection formation in a basidiomycete dikaryon (from *Nonvascular Plants: An Evolutionary Survey* by Scagel, Bandoni, Maze, Rouse, Schofield and Stein. © 1982 by Wadsworth Publishing Co. Reprinted by permission of the publisher).

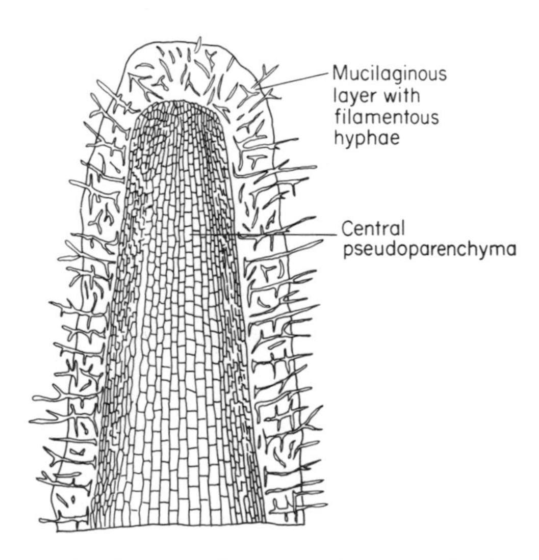

Figure 151. Longitudinal section through a rhizomorph of *Armillaria* (modified after Garret, 1963. *Soil Fungi and Soil Fertility.* © 1963 Pergamon Press Ltd. by permission).

secondary mycelium. The secondary mycelium is responsible for the formation of the tertiary mycelium or basidiocarp (Fig. 152). Nuclear fusion occurs in the basidia in the hymenial layer of the basidiocarp and is followed by meiotic production of four haploid basidiospores.

The rust and smut fungi are classified within the Basidiomycota because they produce basidia and basidiospores. However, the basidia are not produced on a carpophore, and in many other respects these

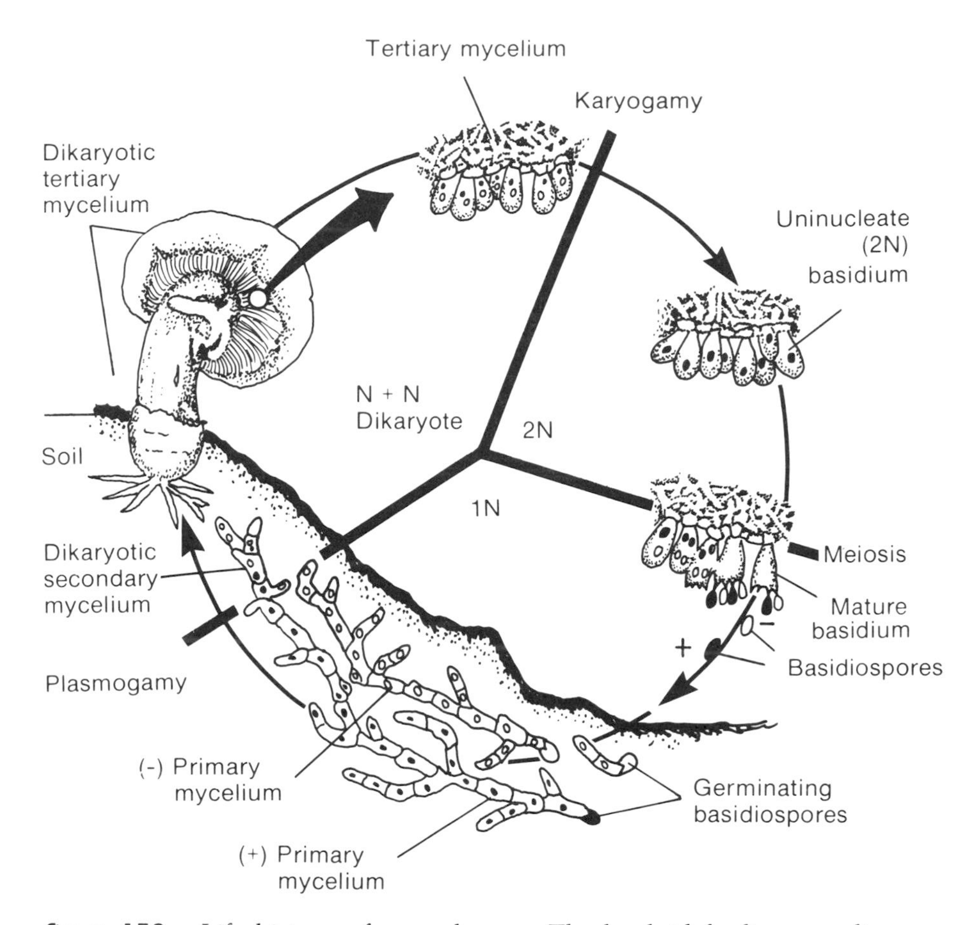

Figure 152. Life history of a mushroom. The haploid feeding mycelium undergoes somatogamy (dimictic) giving rise to an independent dikaryotic mycelium which in turn develops an aerial basidiocarp. The basidiocarp hymenium produces haploid basidiospores that complete the cycle (modified after Pritchard, and Bradt, 1984. *The Biology of Nonvascular Plants.* The C.V. Mosby Co.).

specialized parasites are different from other members of the division. The dikaryons of these fungi give rise to thick-walled resistant spores. When they germinate a septate basidium is formed directly. Each basidium forms four spores in the usual way.

The rust fungi have complicated life histories (not described here) that involve the infection of two different flowering plant host species by different life history phases. One of the hosts is a commercially important crop (wheat, asparagus, beans and many others). The effects of these parasites are devastating.

Apart from the commercial significance of basidiomycetes (positive and negative), it is important to stress their importance in ecological cycles. Members of this division are the most important decomposers of lignin. In addition, basidiomycetes have important mutualistic relations with plant roots which lead to better utilization of mineral nutrients in the soil.

Division Deuteromycota

There are about 10,000 species in this division. They all lack sexual stages. Whenever sexual reproduction is discovered in a member species, it is transferred to the Ascomycota or Basidiomycota, depending on whether asci or basidia are produced. The division Deuteromycota is therefore an artifical construction. The vegetative mycelia share characteristics with other higher fungi (e.g., well-developed hyphal cross walls) which makes classification impossible.

Index

(*continued on next page*)